HiSET MATH 2021 PREPARATION BOOK

High School Equivalency Test Practice Questions
with Math Study Guide

The HiSET® is a trademark of Educational Testing Service (ETS), which is neither affiliated with nor endorses this publication.

HiSET Math 2021 Preparation Book: High School Equivalency Test Practice Questions with Math Study Guide

© COPYRIGHT 2017-2021 Exam SAM Study Aids & Media

ISBN: 978-1-949282-67-2

All rights reserved. No part of this publication may be reproduced, stored in a retrieval system, or transmitted, in any form or by any means, electronic, mechanical, photocopying, recording, or otherwise, without the prior written permission of the copyright owner.

The drawings in this publication are for illustration purposes only. They are not drawn to an exact scale.

Note: The HiSET® is a trademark of Educational Testing Service (ETS), which is neither affiliated with nor endorses this publication.

TABLE OF CONTENTS

How to Use this Publication — i

HiSET Math Practice Test Set 1 – Questions 1 to 80:

Numerical Operations:

Basic Computations with Integers — 1

Multiplying Fractions — 1

Dividing Fractions — 1

Adding Fractions — 1

Lowest Common Denominator — 1

Simplifying Fractions — 2

Subtracting Fractions — 2

Order of Operations (PEMDAS) — 2

Improper Fractions — 2

Percentages — 2

Working with Prices and Profits — 3

Decimals and Decimal Places — 3

Remainders — 3

Operations on Decimal Numbers — 3

Ordering Numbers — 3

Practical Problems — 4

Greatest and Fewest Multiples — 4

Data Interpretation and Probability:

Ratios and Proportions — 4

Probability — 4

Mean — 5

Mode — 5

Median — 5

Range — 5

Variance	5
Standard Deviation	5
Sets	6
Interpreting Data from Charts and Graphs	6

Algebra and Algebraic Patterns:

Roots and Radicals	6
Exponent Laws	7
Simplifying Rational Algebraic Expressions	8
Factoring Polynomials	9
Expanding Polynomials	9
Linear Equations	9
Algebraic and Polynomial Functions	11
Quadratic Equations	12
Linear Inequalities	12
Quadratic Inequalities	13
Systems of Equations	13
Other Algebraic Concepts	13
Logarithmic Functions	14

Measurement and Estimation:

Angle and triangle laws	14
Arcs	15
Area of circles, squares, and triangles	15
Circumference, radius, and diameter	15
Pythagorean theorem	16
30° - 60° - 90° triangles	16
Perimeter	16
Radians	17
Volume of boxes, cones, cylinders, and pyramids	17

Midpoint formula	17
Distance formula	18
Slope and slope-intercept	18
x and y intercepts	18

Trigonometry:

Trigonometric formulas	18
Sine, cosine, and tangent problems	19

Other Topics:

Absolute value	19
Perpendicular lines	20
Function domain and range	20
Leading coefficients and end behavior	20
Graphing Quadratics	21
Transformations	22

HiSET Math Practice Questions 81 to 400:

HiSET Math Practice Test Set 2 – Questions 81 to 160	23
HiSET Math Practice Test Set 3 – Questions 161 to 240	37
HiSET Math Practice Test Set 4 – Questions 241 to 320	50
HiSET Math Practice Test Set 5 – Questions 321 to 400	65

Solutions and Explanations to Questions 1 to 400:

Solutions and Explanations for Practice Test Set 1 – Questions 1 to 80	81
Solutions and Explanations for Practice Test Set 2 – Questions 81 to 160	94
Solutions and Explanations for Practice Test Set 3 – Questions 161 to 240	107
Solutions and Explanations for Practice Test Set 4 – Questions 241 to 320	120
Solutions and Explanations for Practice Test Set 5 – Questions 321 to 400	133

Answer Key to All Questions   144

HOW TO USE THIS PUBLICATION

You will find the problems in this study guide a bit easier to complete if you have knowledge of basic arithmetic operations, such as addition, subtraction, multiplication, and division.

You may feel that you need to review arithmetic and elementary algebra before you try the practice problems in this book.

If so, you should try our free practice exercises in arithmetic and elementary algebra first.

The free review problems can be found at: www.examsam.com

As you work through this study guide, you will notice that practice test questions 1 to 80 provide study tips after each question.

The format of the first set of practice test questions introduces all of the concepts on the exam. This will help you learn the strategies and formulas that you need to answer all of the types of questions on the actual examination.

You can refer back to the formulas and tips introduced in part 1 as you work through the remaining material in the book.

Ideally, you should try to memorize the formulas and tips before you complete the remaining practice test questions in the book.

The solutions and explanations for all of the questions are provided after the 400th question.

The answer key is provided at the end of the book.

HiSET Math Practice Test Set 1 – Questions 1 to 80:

Numerical Operations Problems

1) − (−6) + 2 = ?
A) −8
B) −4
C) 4
D) 8

> Computations with signed numbers are frequently included on the examination. Many of these types of problems will involve integers. Integers are positive and negative whole numbers. Integers cannot have decimals, nor can they be mixed numbers. In other words, they can't contain fractions. One of the most important concepts to remember when working with signed numbers is that two negative signs together make a positive number. So, when you see a number like − (−2) you have to use 2 in your calculation.

2) What is the largest possible product of two even integers whose sum is 30?
A) 14
B) 16
C) 210
D) 224

> You will also see problems that ask you to perform multiplication or division on integers. Some of these problems may ask you to find an integer that meets certain mathematical requirements, like the problem above.

3) $1/5 \times 2/3 = ?$
A) $1/4$
B) $3/8$
C) $2/15$
D) $10/3$

> You will see problems on the exam that ask you to multiply fractions. To multiply fractions, you first need to multiply the numerators from each fraction. Then multiply the denominators. The numerator is the number on the top of each fraction. The denominator is the number on the bottom of the fraction.

4) $\frac{4}{7} \div \frac{2}{3} = ?$
A) $\frac{8}{21}$
B) $\frac{12}{14}$
C) $\frac{6}{8}$
D) $\frac{14}{12}$

> You will also need to know how to divide fractions for the exam. To divide fractions, invert the second fraction by putting the denominator on the top and numerator on the bottom. Then multiply as indicated for the previous problem.

5) $\frac{1}{8} + \frac{3}{16} = ?$
A) $\frac{5}{16}$
B) $\frac{4}{24}$
C) $\frac{16}{5}$
D) $\frac{24}{16}$

> In some fraction problems, you will have to find the lowest common denominator. In other words, before you add or subtract fractions, you have to change them so that the bottom numbers in each fraction are the same. You do this by multiplying the numerator by the same number that you used when multiplying to get the new denominator for the fraction.

6) Simplify: $\frac{12}{14}$

A) $\frac{1}{7}$
B) $\frac{4}{7}$
C) $\frac{7}{6}$
D) $\frac{6}{7}$

> You will also need to know how to simplify fractions for your exam. To simplify fractions, look to see what factors are common to both the numerator and denominator. Factoring is like taking a number apart. So, what numbers can we multiply together to get 12? What numbers can we multiply together to get 14?

7) $4\frac{1}{8} - 3\frac{5}{6} = ?$

A) $-1\frac{1}{2}$
B) $1\frac{17}{24}$
C) $\frac{7}{24}$
D) $\frac{24}{7}$

> Mixed numbers are those that contain a whole number and a fraction. Convert the mixed numbers back to fractions first. Then find the lowest common denominator of the fractions in order to solve the problem.

8) $-5 \times 4 - 6 \div 3 = ?$

A) -22
B) $-\frac{26}{3}$
C) -18
D) $\frac{10}{3}$

9) Express as an improper fraction: $\dfrac{4 \times (5-2)^3 + 6}{7 - 4 \div 2}$

A) $\frac{60}{5}$
B) $\frac{114}{5}$
C) $\frac{114}{1.5}$
D) $\frac{60}{1.5}$

> These two questions test your knowledge of order of operations. The phrase "order of operations" means that you need to know which mathematical operation to do first when you are faced with longer problems. Remember the acronym PEMDAS. "PEMDAS" means that you have to do the mathematical operations in this order: First: Do operations inside **P**arentheses; Second: Do operations with **E**xponents; Third: Do **M**ultiplication and **D**ivision (from left to right); Last: Do **A**ddition and **S**ubtraction (from left to right)

10) The price of a certain book is reduced from $60 to $45 at the end of the semester. By what percent is the price of the book reduced?

A) 15%
B) 20%
C) 25%
D) 33%

> You will have to calculate percentages and decimals on the exam, as well as use percentages and decimals to solve other types of math problems or to create equations. Percentages can be expressed by using the symbol %. They can also be expressed as fractions or decimals. This particular question asks you to perform a calculation in order to determine the percentage discount on an item. Calculate the dollar amount of the reduction by subtracting the sales price from the original price. Then divide the dollar value of the reduction by the original price to get the percentage.

11) A company buys an item at a cost of B and sells it at five times the cost. Which of the following represents the profit made on each item?

A) B
B) 4B
C) 5B
D) 5 − B

You will see questions on the test that ask you solve real-life problems from basic information. Read the problem carefully and then decide which arithmetic operations are needed in order to solve it. The problem tells us that cell phones sell for four times the cost, so "four times" means that we have to multiply. For this problem, profit is calculated by taking the sales price and subtracting the cost.

12) What is 0.96547 rounded to the nearest thousandth?
A) 0.96 B) 0.97 C) 0.966 D) 0.965

13) When 1523.48 is divided by 100, which digit of the resulting number is in the tenths place?
A) 1 B) 2 C) 3 D) 4

These two questions assess your understanding of decimals. Remember that the number after the decimal is in the tenths place, the second number after the decimal is in the hundredths place, and the third number after the decimal is in the thousandths place.

14) What is the remainder when 11 is divided by 3?
A) .66 B) .67 C) 2 D) 3

The remainder is the amount that is left over after you divide into whole numbers. These whole numbers are referred to as factors. So, ask yourself what products can be calculated by multiplying another number by 3: 1 × 3 = 3; 2 × 3 = 6; 3 × 3 = 9; 4 × 3 = 12 and so on.

15) 4.25 + 0.003 + 0.148 = ?
A) 4.401 B) 4.428 C) 5.76 D) 44.01

Line up the decimal points when you add up and remember to carry the 1 where needed.

16) The numbers in the following list are ordered from greatest to least: θ, η, $25/13$, $10/9$, $1/3$
Which of the following could be the value of η?
A) $\sqrt{36}$ B) $25/14$ C) $24/13$ D) 1.91

This problem is asking you to determine missing values from an ordered list of fractions and other numbers. You may find it easier to solve problems like this one if you convert the fractions to decimals.

17) If $7x$ is between 5 and 6, which of the following could be the value of x?
A) $2/3$ B) $3/4$ C) $5/8$ D) $7/8$

To solve the problem, set up an inequality as follows: $5 < 7x < 6$. Then put the fractions from the answer choices in for x in order to solve the problem. When a problem asks you to multiply a whole number by a fraction, multiply the whole number by the numerator and then divide this result by the denominator.

18) The temperature on Saturday was 62° F at 5:00 PM and 38° F at 11:00 PM. If the temperature fell at a constant rate on Saturday, what was the temperature at 9:00 PM?
A) 58° F B) 54° F C) 50° F D) 46° F

This question assesses your knowledge of performing operations on integers. Here, we have to perform the operations of subtraction, multiplication, and division.

19) A painter needs to paint 8 rooms, each of which have a surface area of 2000 square feet. If one bucket of paint covers 900 square feet, what is the fewest number of buckets of paint that must be used to complete all 8 rooms?
A) 3　　　　　　　B) 17　　　　　　　C) 18　　　　　　　D) 19

This is a question that requires you to find the fewest multiples of an item. Be mindful of the words "fewest" and "greatest" in problems like this one, since it will normally be impossible to purchase a fractional part of the item in the question. Therefore, you will need to round your result up or down to the nearest whole number accordingly.

20) Soon Li jogged 3.6 miles in 3/4 of an hour. What was her average jogging speed in miles per hour?
A) 2.7　　　　　　B) 4.0　　　　　　C) 4.2　　　　　　D) 4.8

This problem involves the calculation of miles per hour with fractional parts of hours. To solve the problem, divide the distance traveled by the time in order to get the speed in miles per hour.

Data Interpretation and Probability Problems

21) The ratio of males to females in the senior year class of Carson Heights High School was 6 to 7. If the total number of students in the class is 117, how many males are in the class?
A) 48　　　　　　　B) 54　　　　　　　C) 56　　　　　　　D) 58

Remember that a ratio can be expressed by using the word "to" or by separating the amounts in the subsets with a colon. So, our ratio is expressed as 6 to 7 or 6:7.

22) Aleesha rolls a fair pair of six-sided dice. Each die has values from 1 to 6. She rolls an even number on her first roll. What is the probability that she will roll an odd number on her next roll?

A) $1/2$　　　　　　B) $1/6$　　　　　　C) $2/6$　　　　　　D) $6/11$

This is a question on calculating basic probability. First of all, calculate how many items there are in total in the data set, which is also called the "sample space" or (S). Then reduce the data set if further items are removed. Probability can be expressed as a fraction. The number of items available in the total data set at the time of the draw goes in the denominator. The chance of the desired outcome, which is also referred to as the event or (E), goes in the numerator of the fraction. You can determine the chance of the event by calculating how many items are available in the subset of the desired outcome.

23) A student receives the following scores on his exams during the semester: 89, 65, 75, 68, 82, 74, 86. What is the mean of his scores?

A) 24　　　　　　　B) 74　　　　　　　C) 75　　　　　　　D) 77

The arithmetic mean is the same thing as the arithmetic average. In order to calculate the mean, add up the values of all of the items in the set, and then divide by the number of items in the set.

24) Members of a weight loss group report their individual weight loss to the group leader every week. During the week, the following amounts in pounds were reported: 1, 1, 3, 2, 4, 3, 1, 2, and 1. What is the mode of the weight loss for the group?
A) 1 pound   B) 2 pounds   C) 3 pounds   D) 4 pounds

This is a question on mode. Mode is the value that occurs most frequently in a data set. For example, if 10 students scored 85 on a test, 6 students scored 90, and 4 students scored 80, the mode score is 85.

25) Mark's record of times for the 400 meter freestyle at swim meets this season is: 8.19, 7.59, 8.25, 7.35, and 9.10. What is the median of his times?
A) 7.59   B) 8.19   C) 8.25   D) 8.096

This question is asking you to find the median of a set of numbers. The median is the number that is in the middle of the set when the numbers are in ascending order.

26) A student receives the following scores on her assignments during the term: 98.5, 85.5, 80.0, 97, 93, 92.5, 93, 87, 88, 82. What is the range of her scores?
A) 17.0   B) 18.0   C) 18.5   D) 89.65

This is a question on calculating range. To calculate range, the lowest value in the data set is deducted from the highest value in the data set.

27) Six students in an advanced algebra class got the following grades on a semester test: 99, 98, 74, 69, 87, and 79. What is the variance of these exam scores?
A) 85   B) 125   C) 510   D) 750

The variance measures how far each individual number in a set is from the mean. The variance of a data set is calculated as follows:
Step 1 – Find the arithmetic mean (or average) for the data set.
Step 2 – Calculate the "difference from the mean" for each item in the set. You can calculate the "difference from the mean" by subtracting the mean from each value in the data set.
Step 3 – Square the "difference from the mean" for each item by multiplying the value of the item by itself.
Step 4 – Determine the mean of the individual amounts from step 3 above to calculate the variance.

28) What is the standard deviation for the data set in the question above?
A) 5   B) 125   C) $5\sqrt{5}$ D) 15,625

Standard deviation measures how spread out the data is, compared to the mean or expected value. The standard deviation is calculated by taking the square root of the variance.

29) A = {5, 10, 15, 20, 25}
    B = {4, 8, 12, 16, 20}
    C = {12, 14, 16, 18, 20}

Given sets A, B, and C above, which of the following represents A ∪ (B ∩ C) ?

A) {20}
B) {12, 16, 20}
C) {5, 10, 12, 15, 16, 20, 25}
D) {4, 5, 8, 10, 12, 14, 15, 16, 18, 20, 25}

> The question is asking for the union of A with the intersection of B and C. Since (B ∩ C) is in parentheses, we need to find that intersection first.

30) A zoo has reptiles, birds, quadrupeds, and fish. At the start of the year, they have a total of 1,500 creatures living in the zoo. The pie chart below shows percentages by category for the 1,500 creatures at the start of the year.

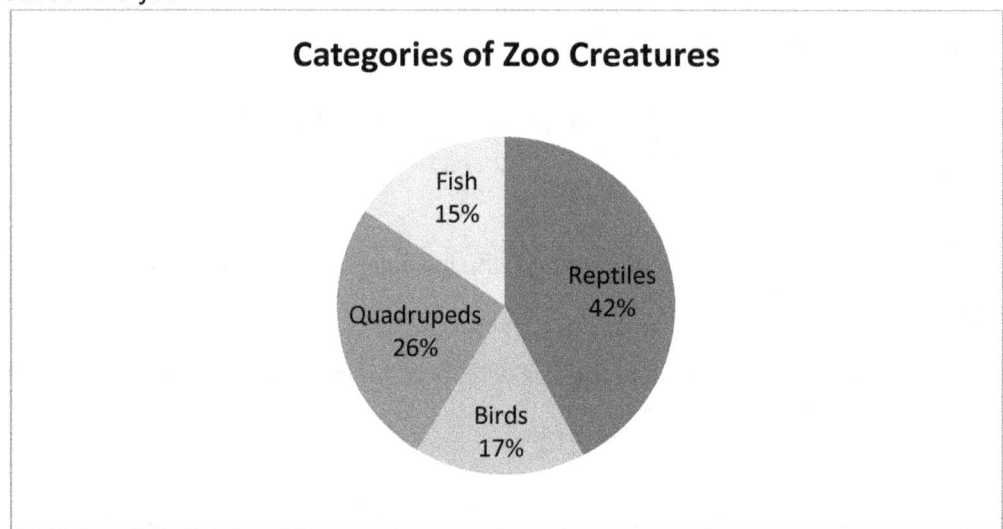

At the end of the year, the zoo still has 1,500 creatures, but reptiles constitute 40%, birds 23%, and quadrupeds 21%. How many more fish were there at the end of the year than at the beginning of the year?

A) 10   B) 11   C) 15   D) 16

> You will see questions on charts and graphs in the quantitative reasoning part of the exam. This question is asking you to interpret a pie chart that shows percentages by category. If you are asked to calculate changes to the data in the categories in the chart, be sure to multiply by the percentages at the beginning of the year and then do a separate calculation using the percentages at the end of the year.

Algebra and Algebraic Patterns

Roots and Radicals

31) Which of the answers below is equal to the following radical expression? $\sqrt{50}$

A) 1 ÷ 50   B) $2\sqrt{25}$ C) $2\sqrt{5}$ D) $5\sqrt{2}$

Step 1: Factor the number inside the square root sign. Step 2: Look to see if any of the factors are perfect squares. In this case, the only factor that is a perfect square is 25. Step 3: Find the square root of 25 then simplify.

32) $\sqrt{36} + 4\sqrt{72} - 2\sqrt{144} = ?$
A) $2\sqrt{36}$
B) $2\sqrt{252}$
C) $18 + 24\sqrt{2}$
D) $-18 + 24\sqrt{2}$

Step 1: Find the common factors that are perfect squares. Step 2: Factor the amounts inside each of the radical signs and simplify.

33) $\sqrt{7} \times \sqrt{11} = ?$
A) $\sqrt{77}$
B) $\sqrt{18}$
C) $7\sqrt{11}$
D) $11\sqrt{7}$

Step 1: Multiply the numbers inside the radical signs. Step 2: Put this product inside a radical symbol for your answer.

34) Express as a rational number: $\sqrt[3]{\dfrac{216}{27}}$
A) 3
B) 2
C) $\dfrac{7}{3}$
D) $\sqrt[3]{2}$

Step 1: Find the cube roots of the numerator and denominator to eliminate the radical. Step 2: Simplify further if possible. The cube root is a number that equals the required product when multiplied by itself two times.

Exponent Laws

35) $7^5 \times 7^3 = ?$
A) 7^8
B) 7^{15}
C) 14^8
D) 49^8

If the base number is the same, you need to add the exponents when multiplying, but keep the base number the same as before.

36) $xy^6 \div xy^3 = ?$
A) xy^{18}
B) xy^3
C) x^2y^3
D) xy^2

If the base number is the same, you need to subtract the exponents when dividing, but keep the base number the same as before.

37) A rocket flies at a speed of 1.7×10^5 miles per hour for 2×10^{-1} hours. How far has this rocket gone?
A) 340,000 miles
B) 34,000 miles
C) 3,400 miles
D) 340 miles

Step 1: Add the exponents to multiply the 10's. Step 2: Multiply the miles per hour by the number of hours to get the distance traveled. Step 3: Then multiply these two results together to solve the problem.

38) $\sqrt{x^{\frac{5}{7}}} = ?$

A) $\frac{5x}{7}$

B) $\left(\sqrt[5]{x}\right)^7$

C) $\left(7\sqrt{x}\right)^5$

D) $\left(\sqrt[7]{x}\right)^5$

Step 1: Put the base number inside the radical sign. Step 2: The denominator of the exponent is the nth root of the radical. Step 3: The numerator is the new exponent.

39) $x^{-5} = ?$

A) $\frac{1}{x^{-5}}$

B) $\frac{1}{x^5}$

C) $-5x$

D) $\frac{1}{5x}$

Step 1: Set up a fraction, where 1 is the numerator. Step 2: Put the term with the exponent in the denominator, but remove the negative sign on the exponent.

40) $62^0 = ?$

A) -62

B) 0

C) 1

D) 62

Any non-zero number raised to the power of zero is equal to 1.

Simplifying Rational Algebraic Expressions

41) $\dfrac{b + \frac{2}{7}}{\frac{1}{b}} = ?$

A) $b^2 + \frac{7}{2}$

B) $2b + \frac{7}{2}$

C) $b^2 + \frac{2b}{7}$

D) $\frac{b}{b + \frac{2}{7}}$

Step 1: When the expression has fractions in both the numerator and denominator, treat the line in the main fraction as the division symbol. Step 2: Invert the fraction that was in the denominator and multiply.

42) $\dfrac{x^2}{x^2 + 2x} + \dfrac{8}{x} = ?$

A) $\dfrac{x + 8x + 16}{x^2 + 2x}$

B) $\dfrac{x^2 + 8}{x^2 + 3x}$

C) $\dfrac{8x^2 + 16x}{x^3}$

D) $\dfrac{x^2 + 8x + 16}{x^2 + 2x}$

Step 1: Find the lowest common denominator. Since x is common to both denominators, we can convert the denominator of the second fraction to the LCD by multiplying the numerator and denominator of the second fraction by $(x + 2)$. Step 2: When you have both fractions in the LCD, add the numerators to solve.

Factoring Polynomials

43) Perform the operation and simplify: $\dfrac{2a^3}{7} \times \dfrac{3}{a^2} = ?$

A) $\dfrac{6a}{7}$ B) $\dfrac{5a^3}{7a^2}$ C) $\dfrac{2a^6}{21}$ D) $\dfrac{21}{2a^6}$

> Step 1: Multiply the numerator of the first fraction by the numerator of the second fraction to get the new numerator. Step 2: Then multiply the denominators. Step 3: Factor out a^2. Step 4: Simplify.

44) $\dfrac{8x+8}{x^4} \div \dfrac{5x+5}{x^2} = ?$

A) $\dfrac{5x^2}{8}$ B) $\dfrac{8}{5x^2}$ C) $\dfrac{3x+3}{x^2}$ D) $\dfrac{x^2+8x+8}{x^4+5x+5}$

> Step 1: Invert and multiply by the second fraction. Step 2: Cancel out (x + 1). Step 3: Cancel out x^2.

Expanding Polynomials

45) Which of the following expressions is equivalent to $(x + 4y)^2$?

A) $2(x + 8y)$ B) $2x + 8y$ C) $x^2 + 8xy^2 + 16y^2$ D) $x^2 + 8xy + 16y^2$

> When expanding polynomials, you should use the FOIL method: First – Outside – Inside – Last.
> We can demonstrate the FOIL method on an example equation as follows:
> $(a + b)(c + d) = (a \times c) + (a \times d) + (b \times c) + (b \times d) = ac + ad + bc + bd$

Linear Equations

46) A mother has noticed that the more sugar her child eats, the more her child sleeps at night. Which of the following graphs best illustrates the relationship between the amount of sugar the child consumes and the child's amount of sleep?

A)

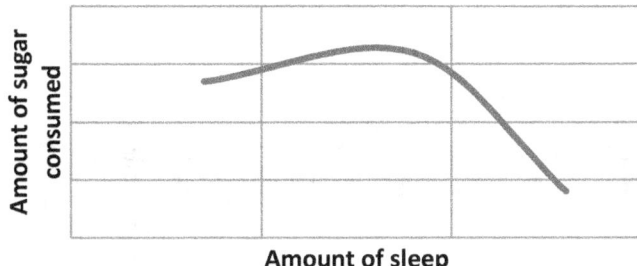

B)

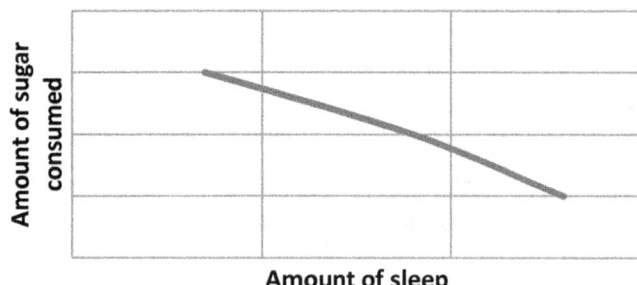

Amount of sleep

C)

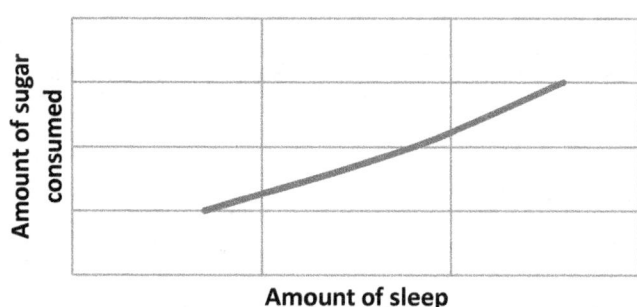

Amount of sleep

D)

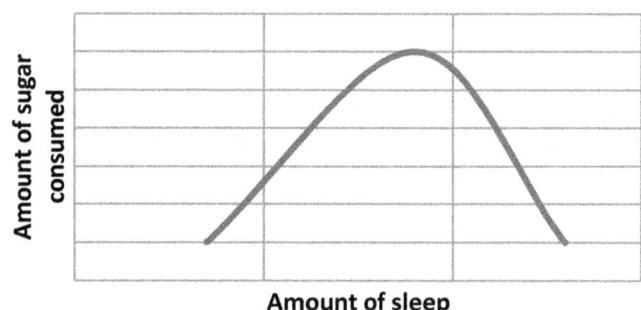

Amount of sleep

You will need to know the difference between positive linear relationships and negative linear relationships for the exam. In a positive linear relationship, an increase in one variable causes an increase in the other variable, meaning that the line will point upwards from left to right.

In a negative linear relationship, an increase in one variable causes a decrease in the other variable, meaning that the line will point downwards from left to right.

Algebraic Functions

47) The graph of a linear equation is shown below. Which one of the tables of values best represents the points on the graph?

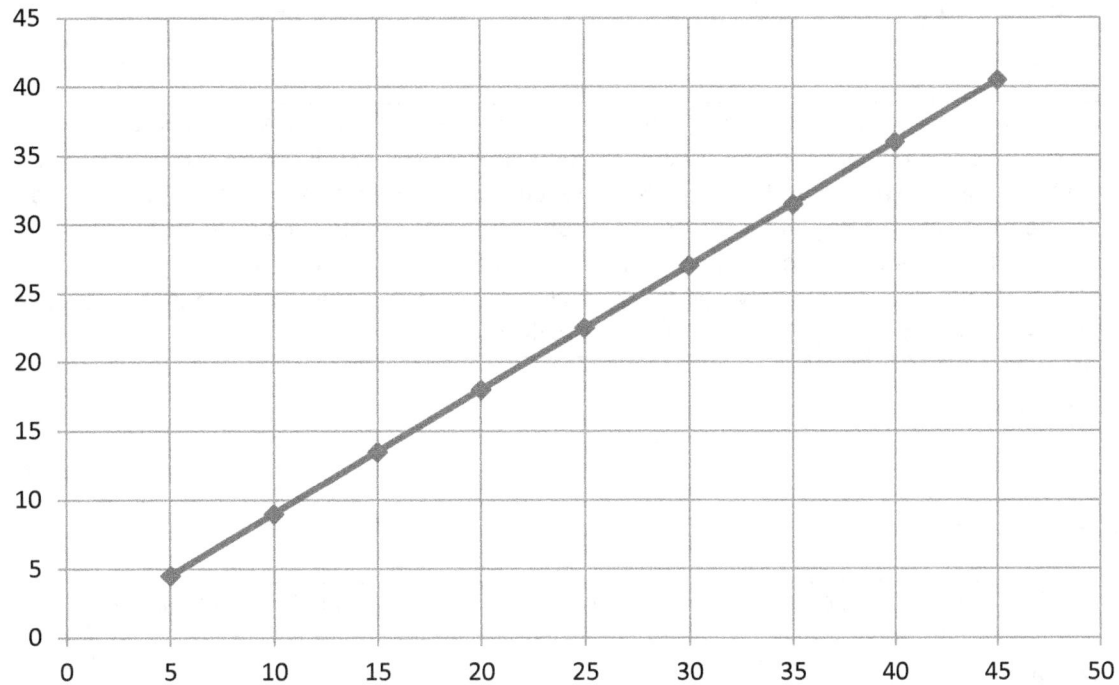

A)

x	y
5	5
10	10
15	15
20	20

B)

x	y
5	4
10	8
15	12
20	16

C)

x	y
5	4.5
10	9
15	13.5
20	18

D)

x	y
5	9
10	13
15	15
20	20

> This is an example of an exam question involving algebraic functions. You will have questions on algebraic, polynomial, exponential, and logarithmic functions on your exam. A function expresses the mathematical relationship between x and y. So, a certain recurring mathematical operation on x will yield the output of y. Step 1: Look carefully at the point that is furthest to the left on the graph. You will be able to eliminate several of the answer choices because they will not state this first coordinate correctly. Step 2: Try to work out the relationship between the coordinates of the first point to those of the next point on the line. Use the horizontal and vertical grid lines on the graph to help you.

Quadratic Equations

48) Simplify: $(x - y)(x + y)$

A) $x^2 - 2xy - y^2$ B) $x^2 + 2xy - y^2$ C) $x^2 + y^2$ D) $x^2 - y^2$

> Use the FOIL method on quadratic equations like this one when the instructions tell you to simplify.

Linear Inequalities

49) $50 - \dfrac{3x}{5} \geq 41$, then $x \leq ?$

A) 15   B) 25   C) 41   D) 50

> Step 1: Isolate the whole numbers to one side of the inequality. Step 2: Get rid of the fraction by multiplying each side by 5. Step 3: Divide to simplify further. Step 4: Isolate the variable to solve.

50) The cost of one wizfit is equal to y. If $x - 2 > 5$ and $y = x - 2$, then the cost of 2 wizfits is greater than which one of the following?

A) $x - 2$ B) $x - 5$ C) $y + 5$ D) 10

> Look to see if the inequality and the equation have any variables or terms in common. In this problem, both the inequality and the equation contain $x - 2$. The cost of one wizfit is represented by y, and y is equal to $x - 2$. So, we can substitute values from the equation to the inequality.

Quadratic Inequalities

51) Solve for x: $x^2 - 9 < 0$

A) $x < -3$ or $x > 3$ B) $x > -3$ or $x < 3$
C) $x < -3$ or $x < 3$ D) $x > -3$ or $x > 3$

For quadratic inequality problems like this one, you need to factor the inequality first. We know that the factors of −9 are: −1 × 9; −3 × 3; 1 × −9. We do not have a term with only the x variable, so we need factors that add up to zero. −3 + 3 = 0. So, try to solve the problem based on these facts. Be sure to check your work when you have found a solution.

Systems of Equations

52) What ordered pair is a solution to the following system of equations?
    x + y = 7
    xy = 12

A) (2, 6)   B) (6, 2)   C) (4, 2)   D) (3, 4)

Step 1: Look at the multiplication equation and find the factors of 12. Step 2: Add the factors in each set together to see if they equal 7 to solve the addition in the first equation.

53) Solve by elimination: $3x + 3y = 15$ and $x + 2y = 8$

A) x = −18 and y = 13   B) x = −2 and y = 3   C) x = 2 and y = 3   D) x = 3 and y = 2

Step 1: Look at the x term of the first equation, which is 3x. In order to eliminate the x variable, we need to multiply the second equation by 3. Step 2: Subtract this result from the first equation to solve.

Other Algebraic Concepts

54) Find the value of the following:

$$\sum_{x=3}^{5} x - 1$$

A) 2   B) 3   C) 4   D) 9

Step 1: You need to perform the operation at the right-hand side of the sigma sign for x = 3. Step 2: We repeat the operation until we use the number at the top of the sigma sign. Step 3: Add the three individual results together to get your answer.

55) If Ð is a special operation defined by (x Ð y) = (30x ÷ 9y) and (3 Ð y) = 10, then y = ?

A) 1   B) 3   C) 9   D) 30

We have the special operation defined as (x Ð y) = (30x ÷ 9y). Looking at the relationship between the left-hand side and the right-hand side of this equation, we can determine the operations that need to be performed on any new equation containing the operation Ð and variables x and y. For our problem, the new equation will be carried out as follows: Operation Ð is division; the number or variable before the special operation symbol is multiplied by 30; and the number or variable after the special operation symbol is multiplied by 9.

Logarithmic Functions

56) $2 = \log_8 64$ is equivalent to which of the following?
A) 2^8 B) 8^2 C) 64^2 D) 64^8

Solve by substituting values into the equation. $x = \log_y Z$ is the same as $y^x = Z$

Measurement and Estimation

57) The triangle in the illustration below is an isosceles triangle. What is the measurement of ∠B?

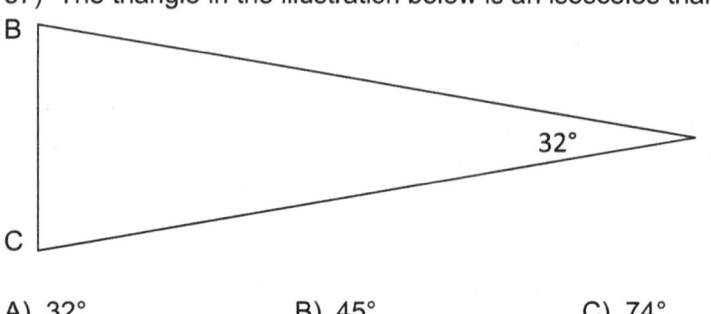

A) 32°   B) 45°   C) 74°   D) 148°

Step 1: Deduct the measurement of angle A from 180° to find out the total degrees of the two other angles. Step 2: Since we have an isosceles triangle, the other two angles are equal in measure.

You should know these principles on angles and triangles for your exam:

The sum of all three angles in any triangle must be equal to 180 degrees.

An isosceles triangle has two equal sides and two equal angles.

An equilateral triangle has three equal sides and three equal angles.

Angles that have the same measurement in degrees are called congruent angles.

Equilateral triangles are sometimes called congruent triangles.

Two angles are supplementary if they add up to 180 degrees. This means that when the two angles are placed together, they will form a straight line on one side.

Two angles are complementary (sometimes called adjacent angles) if they add up to 90 degrees. This means that the two angles will form a right angle.

When two parallel lines are cut by a transversal (a straight line that runs through both of the parallel lines), 4 pairs of opposite (non-adjacent) angles are formed and 4 pairs of corresponding angles are formed. The opposite angles will be equal in measure, and the corresponding angles will also be equal in measure.

A parallelogram is a four-sided figure in which opposite sides are parallel and equal in length. Each angle will have the same measurement as the angle opposite to it, so a parallelogram has two pairs of opposite angles.

The sides of a 30° - 60° - 90° triangle are in the ratio of $1:\sqrt{3}: 2$.

58) The central angle in the circle below is 90° and is subtended by an arc which is 8π centimeters in length. How many centimeters long is the radius of this circle?

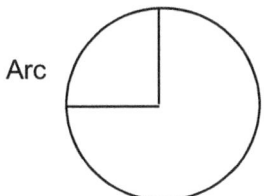
Arc

A) 32   B) 16   C) 8π   D) 8

When working with arcs, you can calculate the radius or diameter of a circle if you have the measurement of a central angle and the length of the arc subtending the central angle. You will also need the formula for circumference: Circumference = π × radius × 2. You can think of arc length as part of the circumference.

59) A field is 100 yards long and 32 yards wide. What is the area of the field in square yards?

A) 160   B) 320   C) 1600   D) 3200

Area of a circle: π × r^2 (radius squared)
Area of a square or rectangle: length × width
Area of a triangle: (base × height) ÷ 2

60) If a circle has a diameter of 6, what is the circumference of the circle?

A) 6π   B) 12π   C) 24π   D) 36π

Diameter = radius × 2
Remember that the formula for the circumference of a circle is π × diameter.

61) If a circle with center (−5, 5) is tangent to the x axis in the standard (x, y) coordinate plane, what is the diameter of the circle?

A) −5   B) −10   C) 5   D) 10

Diameter is the measurement across the entire width of a circle. Diameter is always double the radius. If the center of a circle (x, y) is tangent to the x axis, then both of the following conditions are true: [1] The point of tangency is equal to (x, 0) and [2] The distance between (x, y) and (x, 0) is equal to the radius.

62) XY is 4 inches long and XZ is 5 inches long. What is the area of triangle XYZ?

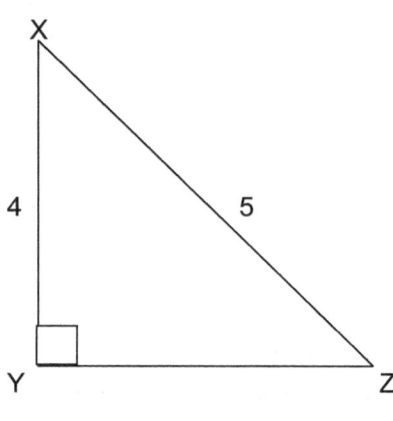

A) 3   B) 5   C) 6   D) 10

Step 1: Use the Pythagorean theorem to find the length of line segment XZ.
Hypotenuse length $C = \sqrt{A^2 + B^2}$
Step 2: Calculate the area of the triangle: (base × height) ÷ 2

63) In the figure below, ∠Y is a right angle and ∠X = 60°. If line segment YZ is 5 units long, then how long is line segment XY?

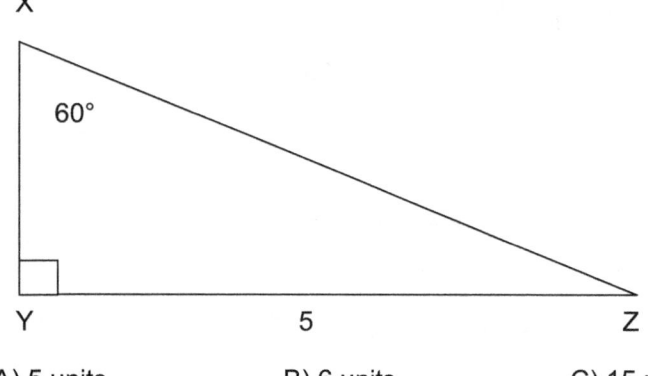

A) 5 units   B) 6 units   C) 15 units   D) $\frac{5}{\sqrt{3}}$ units

Triangle XYZ is a 30° - 60° - 90° triangle.
Using the Pythagorean theorem, its sides are therefore in the ratio of $1: \sqrt{3}: 2$.

64) What is the perimeter of a rectangle that has a length of 17 and a width of 4?

A) 21   B) 34   C) 42   D) 68

In order to calculate the perimeter of squares and rectangles, you need to use the perimeter formula: (length × 2) + (width × 2)

65) If the radius of a circle is 4 and the radians of the subtended angle measure $\pi/4$, what is the length of the arc subtending the central angle?

A) $\pi/4$ B) $\pi/8$ C) π D) 4π

Radians can be illustrated by the diagram and formulas that follow.

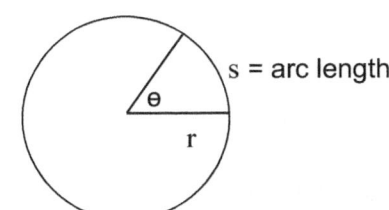

s = arc length

θ = the radians of the subtended angle
s = arc length
r = radius
The following formulas can be used for calculations with radians:
$\theta = s \div r$
$s = r\,\theta$

Also remember these useful formulas.
$\pi \times 2 \times \text{radian} = 360°$
$\pi \times \text{radian} = 180°$
$\pi \div 2 \times \text{radian} = 90°$
$\pi \div 4 \times \text{radian} = 45°$
$\pi \div 6 \times \text{radian} = 30°$

---

66) A cone has a height of 9 inches and a radius at its base of 4 inches. What is the volume of this cone?

A) 13π B) 24π C) 48π D) 144π

Box volume: volume = base × width × height
Cone volume: $(\pi \times \text{radius}^2 \times \text{height}) \div 3$
Cylinder volume: $\pi \times \text{radius}^2 \times \text{height}$
Pyramid volume = (W × L × H) ÷ 3

---

67) If store A is represented by the coordinates (−4, 2) and store B is represented by the coordinates (8, −6), and store A and store B are connected by a line segment, what is the midpoint of this line?

A) (2, 2)   B) (2, −2)   C) (−2, 2)   D) (−2, −2)

The midpoint of two points on a two-dimensional graph is calculated by using the midpoint formula:
$(x_1 + x_2) \div 2 , (y_1 + y_2) \div 2$

---

68) What is the distance between (2, 3) and (6, 7)?

A) 4   B) 16   C) $\sqrt{16}$ D) $\sqrt{32}$

The distance formula is used to calculate the linear distance between two points on a two-dimensional graph. The two points are represented by the coordinates (x_1, y_1) and (x_2, y_2).
$d = \sqrt{(x_2 - x_1)^2 + (y_2 - y_1)^2}$

69) The measurements of a mountain can be placed on a two dimensional linear graph on which $x = 5$ and $y = 315$. If the line crosses the y axis at 15, what is the slope of this mountain?

A) 60   B) 63   C) 300   D) 315

The slope formula: $m = \dfrac{y_2 - y_1}{x_2 - x_1}$

The slope-intercept formula: $y = mx + b$, where m is the slope and b is the y intercept.

70) Find the x and y intercepts of the following equation: $x^2 + 2y^2 = 144$

A) (12, 0) and (0, $\sqrt{72}$)
B) (0, 12) and ($\sqrt{72}$, 0)
C) (0, $\sqrt{72}$) and (0, 12)
D) (12, 0) and ($\sqrt{72}$, 0)

For questions about x and y intercepts, substitute 0 for y in the equation provided. Then substitute 0 for x to solve the problem.

Trigonometry

71) ∠A measures 58° and cos A = sin B. what is the sum of ∠A + ∠B ?

A) 30°   B) 32°   C) 45°   D) 90°

Remember these important trigonometric formulas for the exam:
cos A = sin (90° − A)
sin A = cos (90° − A)
$\cos^2 A + \sin^2 A = 1$
$\cos^2 A = 1 - \sin^2 A$
$\sin^2 A = 1 - \cos^2 A$
tan A = sin A ÷ cos A
tan A × cos A = sin A

72) If x represents a real number, what is the greatest possible value of 4 × cos 2x?

A) 2   B) 3   C) 4   D) 6

The greatest possible value of cosine is 1. Therefore, cos 2x must be less than or equal to 1.

73) Use your knowledge of trigonometric functions as you look at the illustration below: $\sin^2 A = ?$

A) $1 - \cos^2 A$ B) $\sin^2 A - 1$ C) $\tan^2 A$ D) $1 - \tan^2 A$

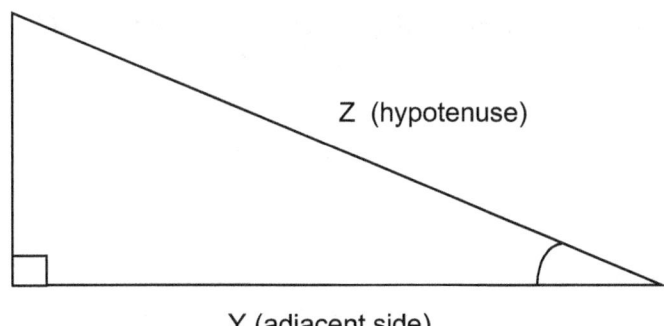

X (opposite side)

Z (hypotenuse)

Angle A

Y (adjacent side)

> To calculate the sine, cosine, and tangent of any given angle A, as in the illustration above:
> $$\sin A = \frac{x}{z} = \frac{opposite}{hypotenuse}$$
> $$\cos A = \frac{y}{z} = \frac{adjacent}{hypotenuse}$$
> $$\tan A = \frac{x}{y} = \frac{opposite}{adjacent}$$

74) If $\cos A = \frac{9}{15}$ and $\tan A = \frac{12}{9}$ then $\sin A = ?$

A) $\frac{12}{9}$ B) $\frac{9}{12}$ C) $\frac{12}{15}$ D) $\frac{15}{12}$

> We know that $\cos A = \frac{y}{z}$ and $\tan A = \frac{x}{y}$. The facts in our problem state: $\cos A = \frac{9}{15}$ and $\tan A = \frac{12}{9}$

Other topics

Absolute Value

75) Which of the following is equivalent to $|-15 + 4|$?

A) $-15 - 4$ B) $-|15 + 4|$ C) $|15| - |4|$ D) $-15 + 4$

> You find the absolute value by making the result of the operation inside the | | signs a positive number. For example, $|-5| = 5$

Perpendicular Lines

76) In the xy plane, line B passes through the origin and is perpendicular to line A. Line A passes through the points (2, –5) and (6, 3). The equation of line B could be which one of the following?

A) $y = -\frac{1}{2}x + 0$

B) $y = \frac{1}{2}x + 0$

C) $y = 2x + 0$

D) $y = -2x + 0$

> To solve questions on perpendicular lines like this one, you first need to find the slope of the line provided in the question. The slopes of perpendicular lines are negative reciprocals of each other. So, to get the negative reciprocal of line A, you first need to invert the integer to make a fraction and then make that fraction negative.

Function Domain and Range

77) Which of the following best describes the range of $y = -3x^2 + 6$?
A) All real numbers.
B) All real numbers except zero.
C) All real positive numbers.
D) All real negative numbers.

> The domain of a function is all possible x values for the function. On the other hand, the range is all of the possible y values for the function. Here, we need to find the range, so we are looking at y values.

Leading Coefficients & End Behavior

78) In the function $f(x) = a(x + 1)(x + 2)$, a is an integer constant. The end behaviour of the graph of $y = f(x)$ is positive for large negative and large positive values of x. Which of the following statements is true with respect to the leading coefficient?
A) The leading coefficient is negative.
B) The leading coefficient is positive.
C) The leading coefficient is an odd number.
D) The leading coefficient is an even number.

> The end behaviour of a graph refers to the values of the polynomial function as it approaches positive or negative infinity. The coefficient that is first in a polynomial as is called a leading coefficient. So, the question is asking us about variable a.

Graphing Quadratics

79) The following graph represents the quadratic function $(x) = ax^2 + bx + c$. In this problem, $y = f(x)$. If d is a constant, how many real solutions exist for the equation $ax^2 + bx + c = dx$?

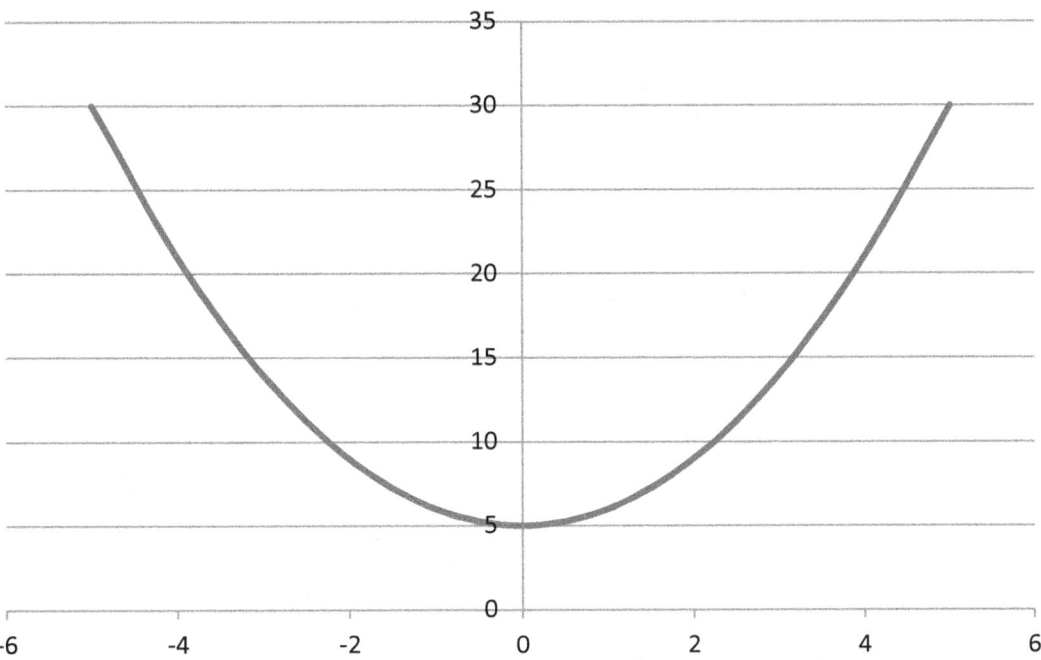

A) 4
B) 2
C) 1
D) 0

As we can see from the graph above, the function $(x) = ax^2 + bx + c$ is graphed as a parabola. The other equation is graphed as a non-vertical line. You need to determine the points of intersection.

Transformations

80)

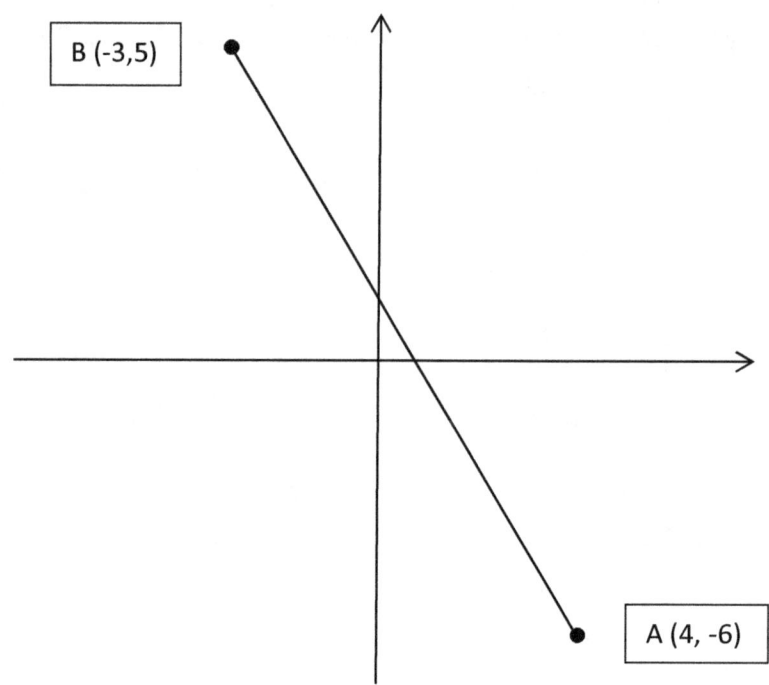

The line in the xy plane above is going to be shifted 2 units to the right and 3 units down. What are the coordinates of point A after the shift?

A) (6, –9)
B) (–1, 2)
C) (–5, 2)
D) (6, 0)

HiSET Math Practice Test Set 2 – Questions 81 to 160:

Numerical Operations Problems

81) A class which has x students. s% of the students have been absent this semester. Which of the following equations represents the number of students who have not been absent this semester?
A) $100(s - x)$
B) $(100\% - s\%) \times x$
C) $(100\% - s\%) \div x$
D) $(1 - s)x$

82) What is 1,594 + 23,786?
A) 24,380
B) 25,380
C) 24.270
D) 25,270

83) A club has 25 members. If each member pays $15 in annual fees, how much money will the club collect in total for the membership fees?
A) $375
B) $355
C) $325
D) $295

84) Which of the following is the greatest?
A) 0.350
B) 0.035
C) 0.053
D) 0.3035

85) A farmer has a field in which cows craze. He is going to buy fence panels to put up a fence along one side of the field. Each panel is 8 feet 6 inches long. He needs 11 panels to cover the entire side of the field. How long is the field?
A) 60 feet 6 inches
B) 72 feet 8 inches
C) 93 feet 6 inches
D) 102 feet 8 inches

86) If the value of x is between 0.0007 and 0.0021, which of the following could be x?
A) 0.0012
B) 0.0006
C) 0.0022
D) 0.022

87) The total funds, represented by variable F, available for P charity projects is represented by the equation F = $500P + $3,700. If the charity has $40,000 available for projects, what is the greatest number of projects that can be completed?
A) 72
B) 73
C) 74
D) 79

88) Which of the following shows the numbers ordered from greatest to least?
A) $-1/3$, $1/7$, 1 , $1/5$
B) $-1/3$, $1/7$, $1/5$, 1
C) $-1/3$, 1 , $1/7$, $1/5$
D) 1 , $1/5$, $1/7$, $-1/3$

89) During each flight, a flight attendant must count the number of passengers on board the aircraft. The morning flight had 52 passengers more than the evening flight, and there were 540 passengers in total on the two flights that day. How many passengers were there on the evening flight?
A) 244
B) 296
C) 488
D) 540

90) A cafeteria serves spaghetti to senior citizens on Fridays. The spaghetti comes prepared in large containers, and each container holds 15 servings of spaghetti. The cafeteria is expecting 82 senior citizens this Friday. What is the least number of containers of spaghetti that the cafeteria will need in order to serve all 82 people?
A) 4 B) 5 C) 6 D) 7

91) A caterpillar travels 10.5 inches in 45 seconds. How far will it travel in 6 minutes?
A) 45 inches B) 63 inches C) 64 inches D) 84 inches

92) Which one of the values will correctly satisfy the following mathematical statement: $2/3 < ? < 7/9$
A) $1/3$ B) $1/5$ C) $2/6$ D) $7/10$

93) Which of the following shows the numbers ordered from least to greatest?
A) 0.2135
   0.3152
   0.0253
   0.0012

B) 0.3152
   0.2135
   0.0253
   0.0012

C) 0.0253
   0.0012
   0.3152
   0.2135

D) 0.0012
   0.0253
   0.2135
   0.3152

94) If $\frac{x}{24}$ is between 8 and 9, which of the following could be the value of x?
A) 190 B) 191 C) 200 D) 217

95) At the beginning of class, $1/5$ of the students leave to go to singing lessons. Then $1/4$ of the remaining students leave to go to the principal's office. If 18 students are then left in the class, how many students were there at the beginning of class?
A) 90 B) 45 C) 30 D) 25

96) A dance academy had 300 students at the beginning of January. It lost 5% of its students during the month. However, 15 new students joined the academy on the last day of the month. If this pattern continues for the next two months, how many students will there be at the academy at the end of March?
A) 285 B) 300 C) 310 D) 315

97) The price of a wool coat is reduced 12.5% at the end of the winter. If the original price of the coat was $120, what will the price be after the reduction?
A) $108.00
B) $107.50
C) $105.70
D) $105.00

98) A motorcycle traveled 38.4 miles in ⁴/₅ of an hour. What was the approximate speed of the motorcycle in miles per hour?
A) 9.6
B) 30.72
C) 48
D) 52

99) A factory that makes microchips produces 20 times as many functioning chips as defective chips. If the factory produced 11,235 chips in total last week, how many of them were defective?
A) 535
B) 561
C) 1,070
D) 10,700

100) A town has recently suffered a flood. The total cost, represented by variable C, which is available to accommodate R number of residents in emergency housing is represented by the equation C = $750R + $2,550. If the town has a total of $55,000 available for emergency housing, what is the greatest number of residents that it can house?
A) 68
B) 69
C) 70
D) 71

Data Interpretation and Probability Problems

101) The pictograph below shows the number of pizzas sold in one day at a local pizzeria. Cheese pizzas sold for $10 each, pepperoni pizzas sold for $12, and the total sales of all three types of pizza was $310. What is the sales price of one vegetable pizza?

Cheese	▼ ▼ ▼
Pepperoni	▼ ▼
Vegetable	▼

Each ▼ represents 5 pizzas.

A) $5
B) $8
C) $9
D) $10

102) The ages of 5 siblings are: 2, 5, 7, 12, and x. If the mean age of the 5 siblings is 8 years old, what is the age (x) of the 5th sibling?
A) 8
B) 10
C) 12
D) 14

103) Sam rolls a fair pair of six-sided dice. One of the die is black and the other is red. Each die has values from 1 to 6. What is the probability that Sam will roll a 4 on the red die and a 5 on the black die?
A) $1/36$
B) $2/36$
C) $1/12$
D) $2/12$

104) In an athletic competition, the maximum possible amount of points was 25 points per participant. The scores for 15 different participants are displayed in the graph below. What was the median score for the 15 participants?

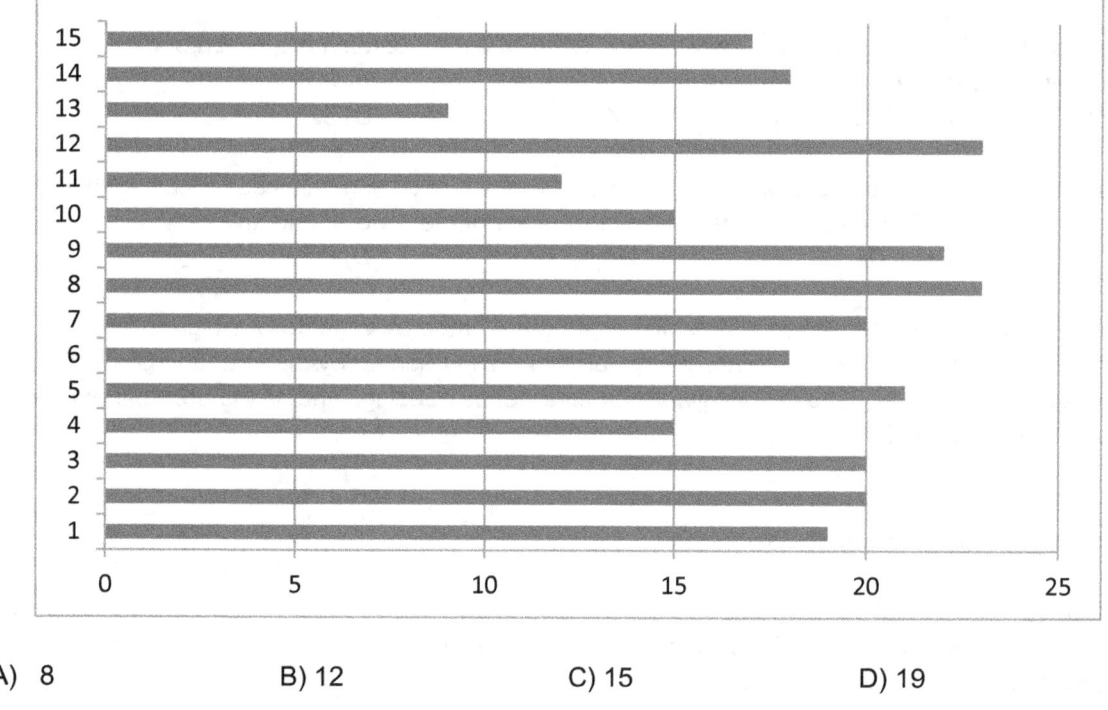

A) 8   B) 12   C) 15   D) 19

105) The table below shows information on disease by different types. Approximately how many cancer and leukemia patients have not survived?

Disease or Complication	Percentage of patients of this disease or complication type that have survived and total number of patients
Cardiopulmonary and vascular	82% (602,000)
HIV/AIDS	73% (215,000)
Diabetes	89% (793,000)
Cancer and leukemia	48% (231,000)
Premature birth complications	64% (68,000)

A) 24,480
B) 110,880
C) 120,120
D) 231,000

106) The combined total of sales for all three of the companies was greatest during which month of the year?

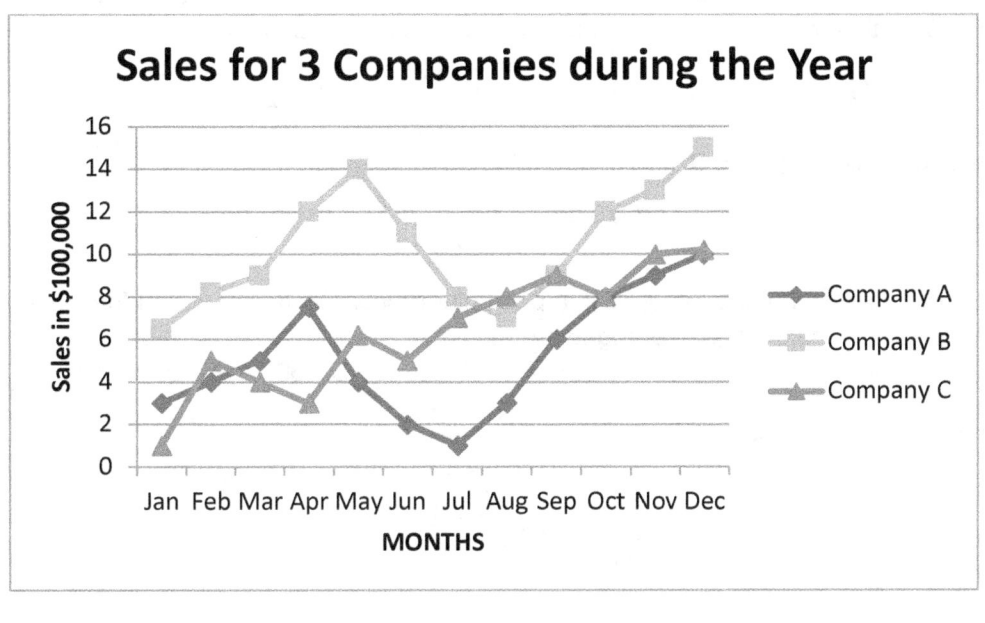

A)   December   B)   November   C)   May   D)   April

107) Calculate the total amount of rainfall in inches for the county that had the least amount of rainfall for all four months in total.

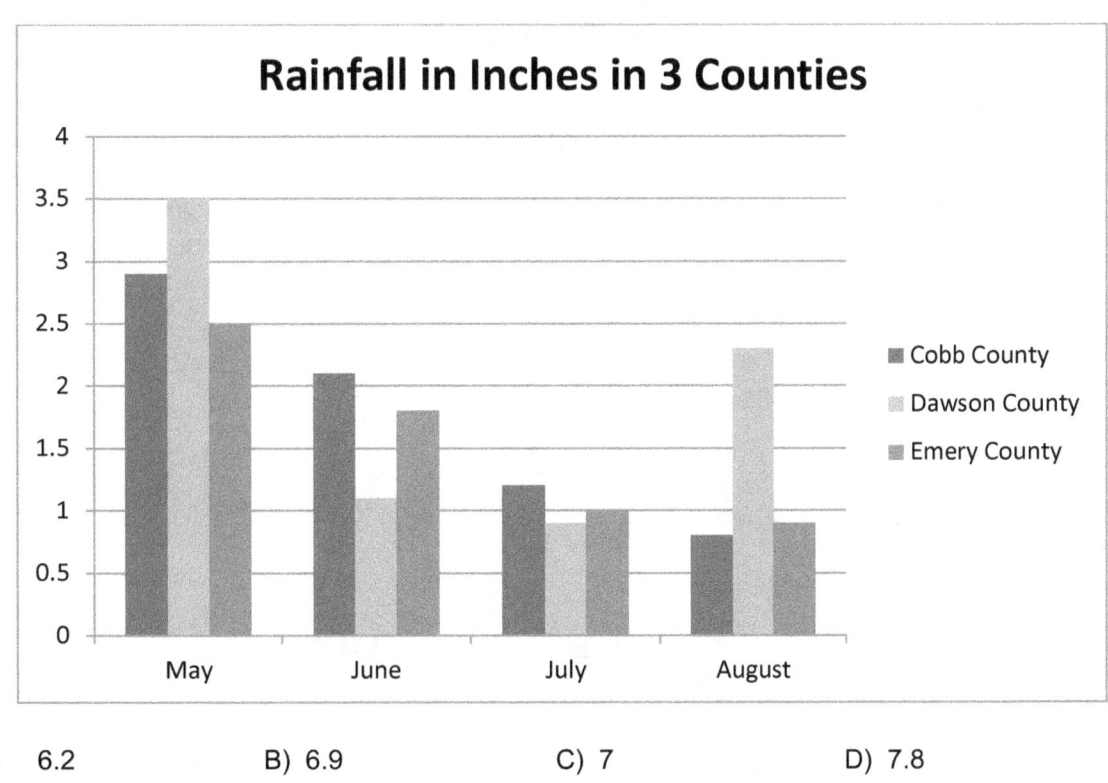

A)   6.2   B)   6.9   C)   7   D)   7.8

108) An owner of a carnival attraction draws teddy bears out of a bag at random to give to prize winners. She has 10 brown teddy bears, 8 white teddy bears, 4 black teddy bears, and 2 pink teddy bears when she opens the attraction at the start of the day. The first prize winner of the day receives a brown teddy bear. What is the probability that the second prize winner will receive a pink teddy bear?

A) $1/24$ B) $1/23$ C) $2/24$ D) $2/23$

109) Find the median of the following: 2.5, 9.4, 3.1, 1.7, 3.2, 8.2, 4.5, 6.4, 7.8

A) 3.2   B) 4.5   C) 5.2   D) 6.4

110) The ratio of bags of apples to bags of oranges in a particular grocery store is 2 to 3. If there are 44 bags of apples in the store, how many bags of oranges are there?
A) 33   B) 48   C) 55   D) 66

Algebra, Measurement, and Estimation

111) Solve for x: $x^2 - 5x + 6 \leq 0$
A) $2 \geq x \geq 3$
B) $2 \leq x \leq 3$
C) $x < -3$ or $x < 2$
D) $x > -2$ or $x > 3$

112) $x^2 + xy - y = 254$ and $x = 12$. What is the value of y?
A) 110   B) 10   C) 11   D) 12

113) $(3x + y)(x - 5y) = ?$
A) $3x^2 - 14xy - 5y^2$
B) $3x^2 - 14xy + 5y^2$
C) $3x^2 + 14xy - 5y^2$
D) $3x^2 + 14xy + 5y^2$

114) Factor: $9x^3 - 3x$
A) $3x(3x^2 - 1)$ B) $3x(3x - 1)$ C) $3x(x^2 - 1)$ D) $3x(x - 3)$

115) The perimeter of the square shown below is 24 units. What is the length of line segment AB?

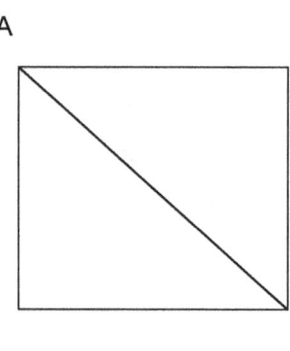

A) $\sqrt{24}$ B) $\sqrt{36}$ C) $\sqrt{72}$ D) 6

116) Which of the following statements is true with respect to the lined graph below?

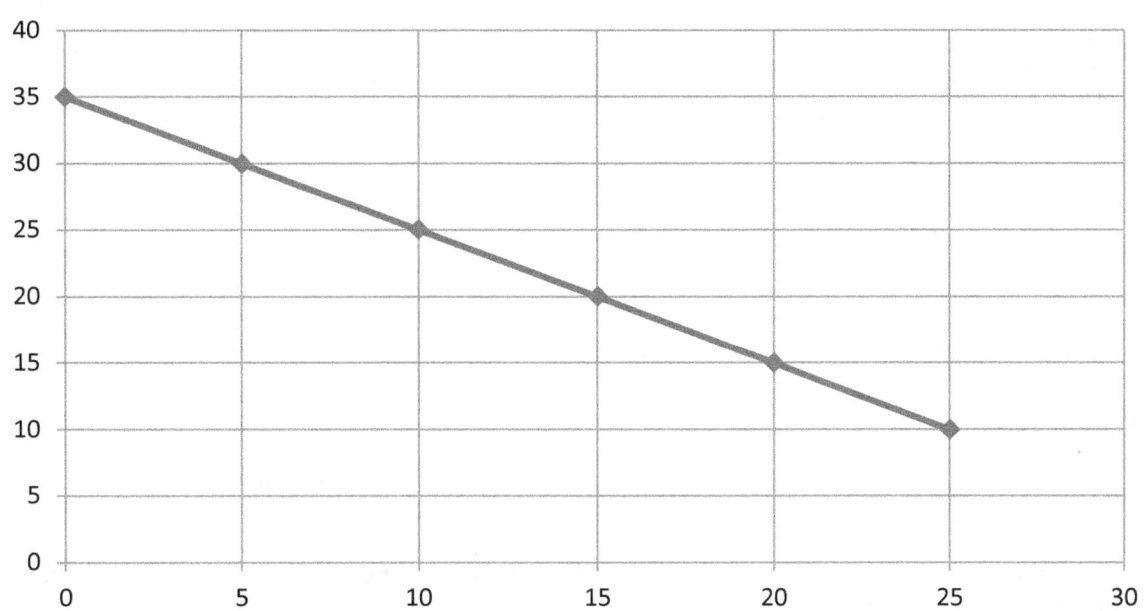

A) The line has a slope of −1 and contains point (20, 15).
B) The line has a slope of 1 and contains point (20, 15).
C) The line has a slope of −1 and contains point (15, 20).
D) The line has a slope of 1 and contains point (15, 20).

117) Simplify: $\sqrt{15} + 3\sqrt{15}$
A) 45
B) $4\sqrt{15}$
C) $2\sqrt{15}$
D) $3\sqrt{30}$

118) $\sqrt{5} \times \sqrt{3}$ = ?
A) 15
B) $\sqrt{8}$
C) $\sqrt{15}$
D) $5\sqrt{3}$

119) Which of the following expressions is equivalent to $\frac{x}{5} + \frac{y}{2}$?

A) $\frac{x+y}{7}$
B) $\frac{2x+5y}{10}$
C) $\frac{5x+2y}{10}$
D) $\frac{2x+5y}{7}$

120) What equation represents the slope-intercept formula for the following data?
Through (4, 5); $m = -\frac{3}{5}$

A) $y = -\frac{3}{5}x + 5$
B) $y = -\frac{12}{5}x - 5$
C) $y = -\frac{3}{5}x - \frac{37}{5}$
D) $y = -\frac{3}{5}x + \frac{37}{5}$

121) Which of the following is equivalent to: $\frac{x^2 + 5x + 4}{x^2 + 6x + 5} \times \frac{16}{x + 5}$?

A) $\frac{16x + 64}{x + 5}$
B) $\frac{x + 20}{x + 5}$
C) $\frac{x + 20}{x^2 + 10x + 25}$
D) $\frac{16x + 64}{x^2 + 10x + 25}$

122) Which of the following is equivalent to: $\dfrac{8x-8}{x} \div \dfrac{3x-3}{6x^2}$?

A) $\dfrac{3x^2-3x}{48x^3-48x^2}$
B) $\dfrac{5x-5}{6x^2}$
C) $\dfrac{8(x-1)\times 6x^2}{x \times 3(x-1)}$
D) $16x$

123) $(25x)^0 = ?$

A) 0
B) 5
C) 1
D) 25

124) $4^{11} \times 4^8 = ?$

A) 16^{19}
B) 4^{19}
C) 8^{19}
D) 4^{88}

125) $\sqrt{8x^4} \cdot \sqrt{32x^6} = ?$

A) $8\sqrt{32x^{10}}$
B) $16x^{10}$
C) $16x^5$
D) $256x^{10}$

126) The length of a box is 20 centimeters (cm), the width is 15cm, and the height is 25cm, what is the volume of the box in centimeters?

A) 150
B) 300
C) 750
D) 7500

127) If one leg of a triangle is 5cm and the other leg is 12cm, what is the measurement of the hypotenuse of the triangle?

A) $5\sqrt{12}$cm
B) $12\sqrt{5}$cm
C) $\sqrt{17}$cm
D) 13 cm

128) In the figure below, A and B are parallel lines, and line C is a transversal crossing both A and B. Which three angles are equal in measure?

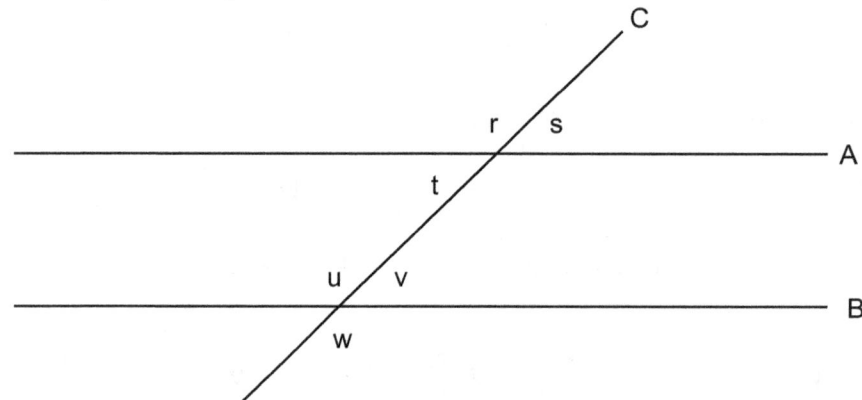

A) ∠r, ∠s, ∠t
B) ∠r, ∠u, ∠v
C) ∠r, ∠u, ∠w
D) ∠r, ∠t, ∠u

129) Express in degrees: $\dfrac{3}{18}\pi$

A) 30°
B) 12°
C) 5°
D) 3°

130) Which of the following dimensions would be needed in order to find the area of the figure?

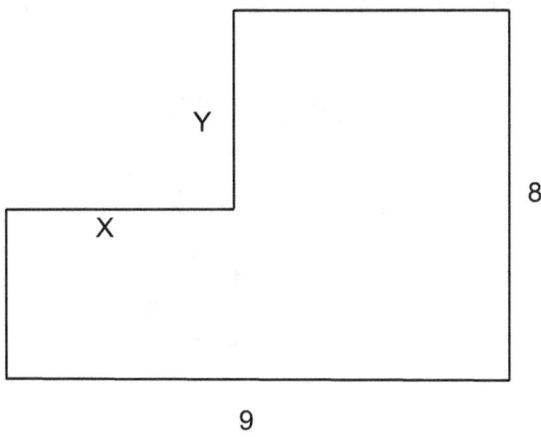

A) X only   B) Y only   C) Both X and Y   D) Either X or Y

131) The figure below shows a right triangular prism. Side AB measures 3.5 units, side AC measures 4 units, and side BD measures 5 units. What amount below best approximates the total surface area of this triangular prism in square units?

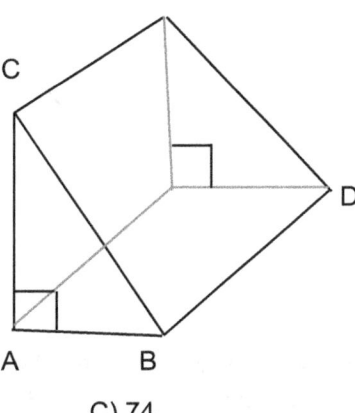

A) 66.5   B) 72.85   C) 74   D) 78.00

132) The base (B) of the cylinder in the illustration shown below is at a right angle to its sides. The radius (R) of the base of cylinder measures 5 centimeters. The height of the cylinder is 14 centimeters. What is the volume of the cylinder?

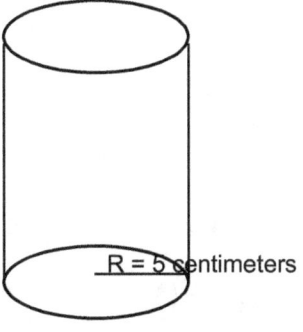

A) 60π   B) 140π   C) 350π   D) 700π

133) Circle 1 and circle 2 are two concentric circles with radii of $R_1 = 1$ and $R_2 = 2.4$ as shown in the illustration below. Line L forms the diameter of the circles. What is the area of the lined part of the illustration?

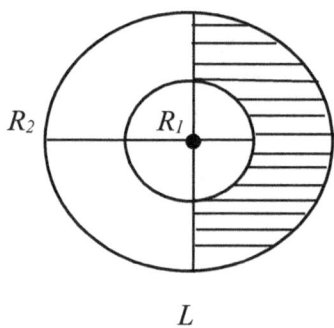

A) 0.7π B) 1.4π C) 2π D) 2.38π

134) AB and CD are parallel and lengths are provided in units. What is the area of trapezoid ABCD in square units?

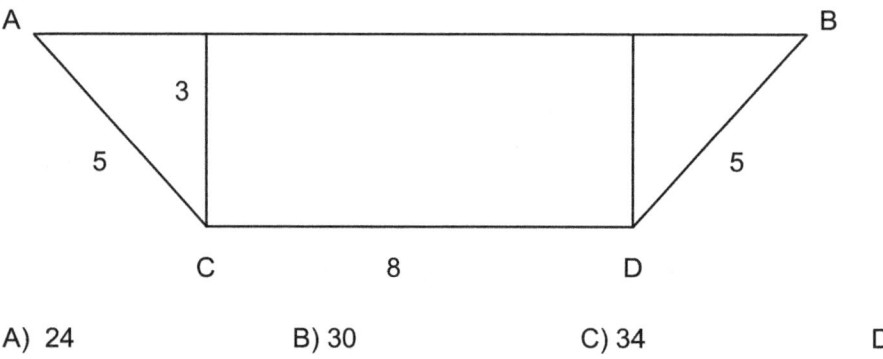

A) 24 B) 30 C) 34 D) 36

135) In the figure below, the circle centered at B is internally tangent to the circle centered at A. The length of line segment AB, which represents the radius of circle A, is 3 units and the smaller circle passes through the center of the larger circle. If the area of the smaller circle is removed from the larger circle, what is the remaining area of the larger circle?

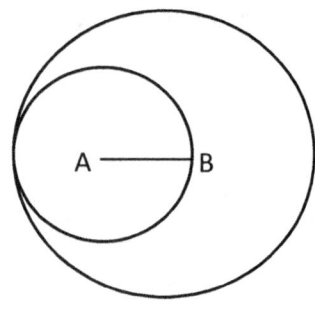

A) 3π B) 6π C) 9π D) 27π

136) A circle has a radius of 12. What is the circumference of the circle?

A) π/12 B) π/24 C) 24π D) 144π

137) Circle M has a radius of 8, and the area of circle M is 39π greater than the area of circle N. What is the radius of circle N?
A) 5π B) 5 C) 6 D) 6.5

138) State the x and y intercepts that fall on the straight line represented by the equation: y = x + 14
A) (–14, 0) and (0, 14)
B) (0, 14) and (0, –14)
C) (14, 0) and (0, –14)
D) (0, –14) and (14, 0)

139) Find the midpoint of the line segment that connects the points (5, 2) and (7, 4).
A) (6, 3) B) (3, 6) C) (3.5, 5.5) D) (12, 6)

140) Express the following as a logarithmic function: $6^4 = 1296$
A) log1296 B) log12.96 × 10^2 C) $6 = \log_4 1296$ D) $4 = \log_6 1296$

141) The graph below illustrates which of the following functions?

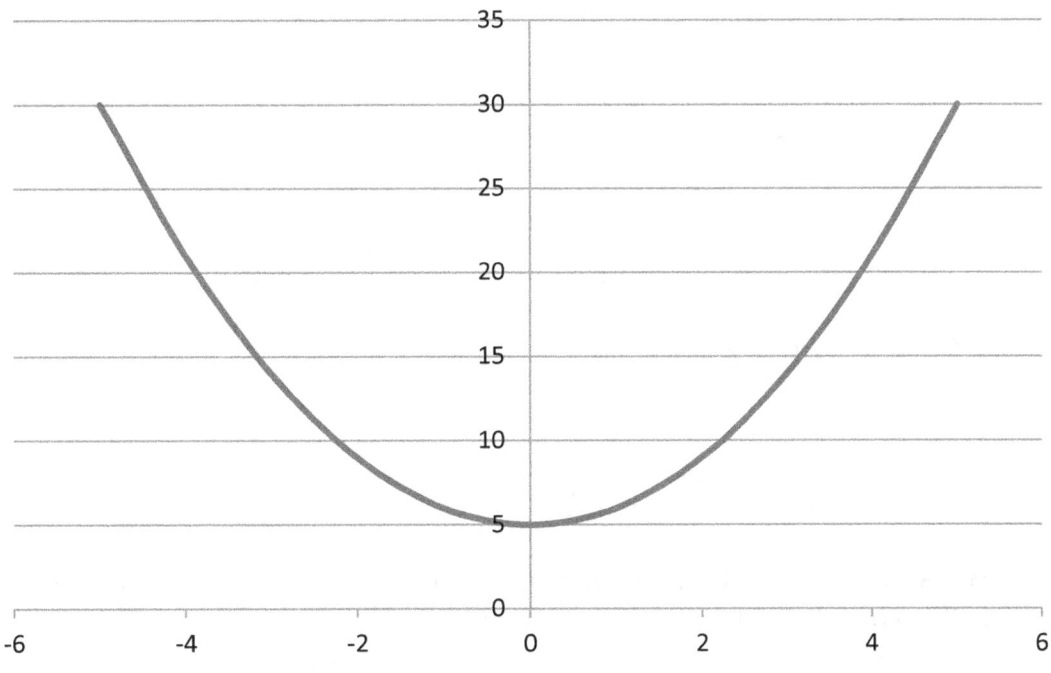

A) $f(x) = x + 5$ B) $f(x) = x - 5$ C) $f(x) = x^2 + 5$ D) $f(x) = x^2 - 5$

142) What is the value of $f_1(2)$ where $f_1(x) = 5^x$?
A) 2^5 B) 10 C) 25 D) 25^2

143) $(-4)^{-3} = ?$
A) –64 B) $-\frac{1}{64}$ C) $\frac{1}{64}$ D) 64

144) In the right triangle below, the length of AC is 10 units and the length of BC is 8 units. What is the tangent of ∠A ?

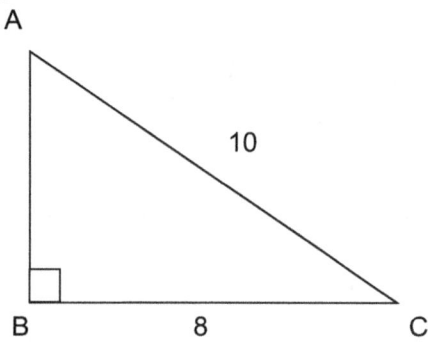

A) 3/4   B) 4/3   C) 6   D) 3/5

145) In the figure below, the length of XZ is 12 units, sin 30° = 0.5, cos 30° = .86603, and tan 30° = 0.57735. Approximately how many units long is XY ?

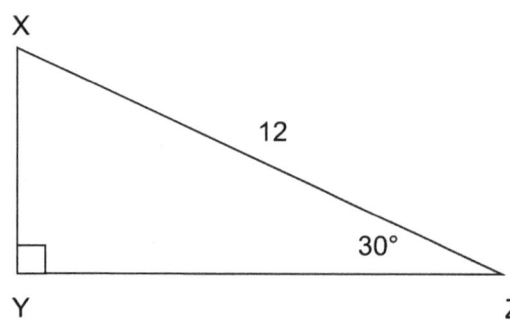

A) 5   B) 5.7735   C) 6   D) 8.6603

146) If cos A = b/c and sin A = a/c then tan A = ?

A) c/a B) c/b C) ab/c D) a/b

147) If cos A = 0.30902 and sin A = 0.95106, then tan A = ?

A) 3.07767   B) 0.307767   C) 0.32492   D) 0.04894

148) A plumber charges $100 per job, plus $25 per hour worked. He is going to do 5 jobs this month. He will earn a total of $4,000. How many hours will he work this month?

A) 32   B) 40   C) 140   D) 160

149) $(2 + \sqrt{6})^2 = ?$

A) 8   B) $8 + 2\sqrt{6}$ C) $8 + 4\sqrt{6}$ D) $10 + 4\sqrt{6}$

150) The cosine of angle Z is 0.78801075 and the tangent of angle Z is 0.7812856. What is the sine of angle Z?

A) 1 − 0.61566148   B) 0.61566148   C) 0.7812856   D) 0.78801075

151) In the xy plane, a line crosses the y axis at point (0, 4) and passes through point (6, 8). Which one of the following could be an equation of the line?

A) $y = \frac{3}{2}x + 4$ B) $y = \frac{2}{3}x + 0$ C) $y = \frac{3}{2}x + 0$ D) $y = \frac{2}{3}x + 4$

152) If $\log_3(x + 2) = 4$, then $x =$?
A) 66   B) 79   C) 81   D) 83

153) What is the domain of $f(x) = x \div (x - 2)$?
A) {2}
B) All real numbers.
C) All real numbers except 2.
D) All real numbers except 0 and 2.

154) The figure in the xy plane below is going to be moved 7 units to the right and 6 units down. What will the coordinates of point C be after the shift?

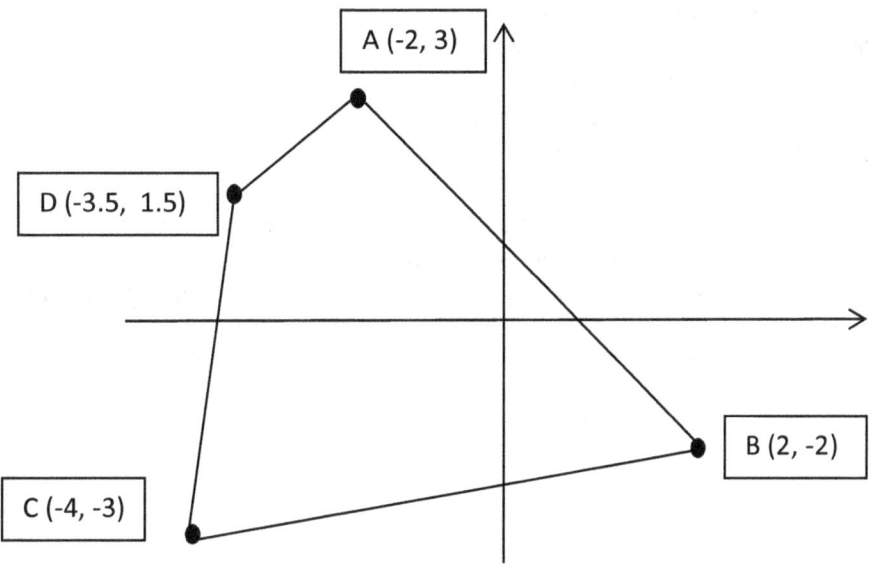

A) (3, 3)   B) (9, –8)   C) (3, –9)   D) (–11, –9)

155) Which of the following best describes the range of $y = x^2 + 36$?
A) All real numbers.
B) All real positive numbers greater than or equal to 36.
C) All real negative numbers less than or equal to –36.
D) All real numbers greater than or equal to 36.

156) Mustafa bought 4 quarts of cranberry juice for $3 per quart and x quarts of orange juice for $2 per quart. The average cost of both drinks was ($12 + $2x) ÷ (4 + x). If the equation ($12 + $2x) ÷ (4 + x) is graphed in the xy plane, what quantity will be represented at the y intercept of the graph?
A) 5   B) 4   C) 3   D) 2

157) Find the value of the following:

$$\sum_{x=3}^{6} x^2 + 5$$

A) 14   B) 55   C) 65   D) 106

158) Toby is going to buy a car. The total purchase price of the car, including interest, is represented by variable C. He will pay D dollars immediately, and then he will make equal payments (P) each month for a certain number of months (M). Which equation below represents the amount of his monthly payment (P)?

A) $\frac{C-D}{M}$ B) $\frac{C}{M} - D$ C) $\frac{M}{C-D}$ D) $D - \frac{C}{M}$

159) There are three boys in a family, named Alex, Burt, and Zander. Alex is twice as old as Burt, and Burt is one year older than three times the age of Zander. Which of the following statements best describes the relationship between the ages of the boys?
A) Alex is 4 years older than 6 times the age of Zander.
B) Alex is 2 years older than 6 times the age of Zander.
C) Alex is 4 years older than 3 times the age of Zander.
D) Alex is 2 years older than 3 times the age of Zander.

160) The price of a sofa at a local furniture store was x dollars on Wednesday this week. On Thursday, the price of the sofa was reduced by 10% of Wednesday's price. On Friday, the price of the sofa was reduced again by 15% of Thursday's price. Which of the following expressions can be used to calculate the price of the sofa on Friday?
A) (0.75)x   B) (0.10)(0.15)x   C) (0.10)(0.85)x   D) (0.90)(0.85)x

HiSET Math Practice Test Set 3 – Questions 161 to 240

Numerical Operations Problems

161) The numbers in the following list are ordered from least to greatest: α, $2/7$, $8/9$, 1.35, $11/3$, μ
Which of the following could be the value of μ?
A) 3.5 B) $10/4$ C) 4.1 D) $1/6$

162) $82 + 9 \div 3 - 5 = ?$
A) −40.50 B) 40.50 C) 80.00 D) 85.33

163) $52 + 6 \times 3 - 48 = ?$
A) 22 B) 82 C) 126 D) 322

164) Convert the following to a decimal: $4/16$
A) 0.0025 B) 0.025 C) 0.25 D) 0.40

165) 90 is 30 percent of what number?
A) 27 B) 120 C) 0.0375 D) 300

166) $6\frac{3}{4} - 2\frac{1}{2} = ?$
A) $4\frac{1}{4}$ B) $4\frac{3}{8}$ C) $4\frac{5}{8}$ D) $4\frac{6}{8}$

167) $9 \times 6 + 42 \div 6 = ?$
A) 8 B) 16 C) 27 D) 61

168) The local Boy Scouts has 31 members. If each member contributes 12 cans of food for a food drive, how many cans of food are contributed in total?
A) 472 B) 372 C) 132 D) 43

169) $1/8 \div 4/3 = ?$
A) $1/6$ B) $32/3$ C) $3/24$ D) $3/32$

170) A group of friends are trying to lose weight. Person A lost $14\frac{3}{4}$ pounds. Person B lost $20\frac{1}{5}$ pounds. Person C lost 36.35 pounds. What is the total weight loss for the group?
A) 70.475 B) 71.05 C) 71.15 D) 71.30

171) Convert the following fraction to decimal format: $5/50$
A) 0.0010 B) 0.0100 C) 0.1000 D) 0.0500

172) What is the remainder when 600 is divided by 9?
A) 0.66 B) 0.67 C) 7 D) 6

173) $3^1/_2 - 2^3/_5 = ?$
A) $^9/_{10}$
B) $1^1/_{10}$
C) $1^1/_3$
D) $1^2/_3$

174) $^1/_6 + (^1/_2 \div ^3/_8) - (^1/_3 \times ^3/_2) = ?$
A) $^{23}/_6$
B) 1
C) 2
D) $^1/_{10}$

175) Mary needs to get $650 in donations. So far, she has obtained 80% of the money she needs. How much money does she still need?
A) $8.19
B) $13.00
C) $32.50
D) $130.00

176) The Abdul family is shopping at a superstore. They buy product A and product B. Product A costs $5 each, and product B costs $8 each. They buy 4 of product A. They also buy a certain quantity of product B. The total value of their purchase is $60. How many units of product B did they buy?
A) 4
B) 5
C) 6
D) 8

177) A hockey team had 50 games this season and lost 20 percent of them. How many games did the team win?
A) 8
B) 10
C) 20
D) 40

178) Jonathan can run 3 miles in 25 minutes. If he maintains this pace, how long will it take him to run 12 miles?
A) 1 hour and 15 minutes
B) 1 hour and 40 minutes
C) 1 hour and 45 minutes
D) 3 hours

179) Mrs. Johnson is going to give candy to the students in her class. The first bag of candy that she has contains 43 pieces. The second contains 28 pieces, and the third contains 31 pieces. If there are 34 students in Mrs. Johnson's class, and the candy is divided equally among all of the students, how many pieces of candy will each student receive?
A) 3 pieces
B) 4 pieces
C) 5 pieces
D) 51 pieces

180) Use the table below to answer the following question:

Sunday	Monday	Tuesday	Wednesday	Thursday	Friday	Saturday
−10°F	−9°F	1°F	6°F	8°F	13°F	12°F

The weather forecast for the coming week is given in the table above. What is the difference between the highest and lowest forecasted temperatures for the week?
A) −2°F
B) −3°F
C) 3°F
D) 23°F

Data Interpretation and Probability Problems

181) Find the value of x that solves the following proportion: $9/6 = x/10$
A) 1.5   B) 15   C) .67   D) 67

182) In a high school, 17 out of every 20 students participate in a sport. If there are 800 students at the high school, what is the total number of students that participate in a sport?
A) 120 students   B) 640 students   C) 680 students   D) 776 students

183) What is the mode of the numbers in the following list? 1.6, 2.9, 4.5, 2.5, 2.5, 5.1, 5.4
A) 3.5   B) 3.1   C) 3.0   D) 2.5

184) There are 10 cars in a parking lot. Nine of the cars are 2, 3, 4, 5, 6, 7, 9, 10, and 12 years old, respectively. If the average age of the 10 cars is 6 years old, how old is the 10th car?
A) 1 year old   B) 2 years old   C) 3 years old   D) 4 years old

185) An employee at the Department of Motor Vehicles wanted to find the mean of the ten driving theory tests they administered this morning. However, the employee divided the total points from the ten tests by 8, which gave him an erroneous result of 78. What is the correct mean of the ten tests?
A) 97.5   B) 70   C) 62.4   D) 52

186) According to the graph, the two highest categories accounted for what percentage of use in total?

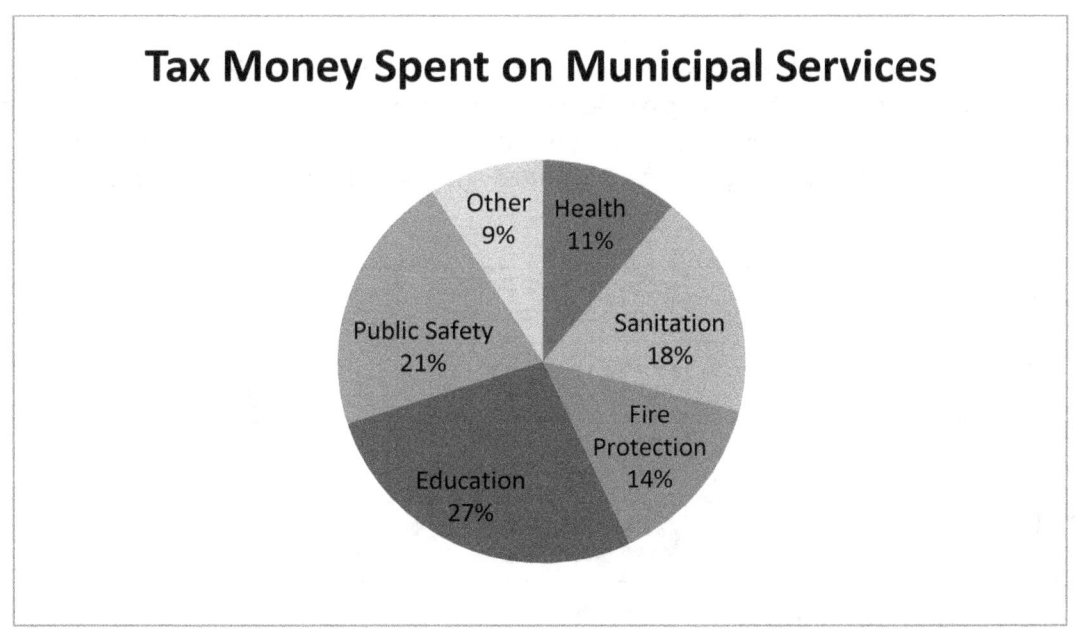

A) 32%   B) 41%   C) 48%   D) 52%

187) The students at Lyndon High School have been asked about their plans to attend the Homecoming Dance. The chart below shows the responses of each grade level by percentages. Which figure below best approximates the percentage of the total number of students from all four grades who will attend the dance? Note that each grade level has roughly the same number of students.

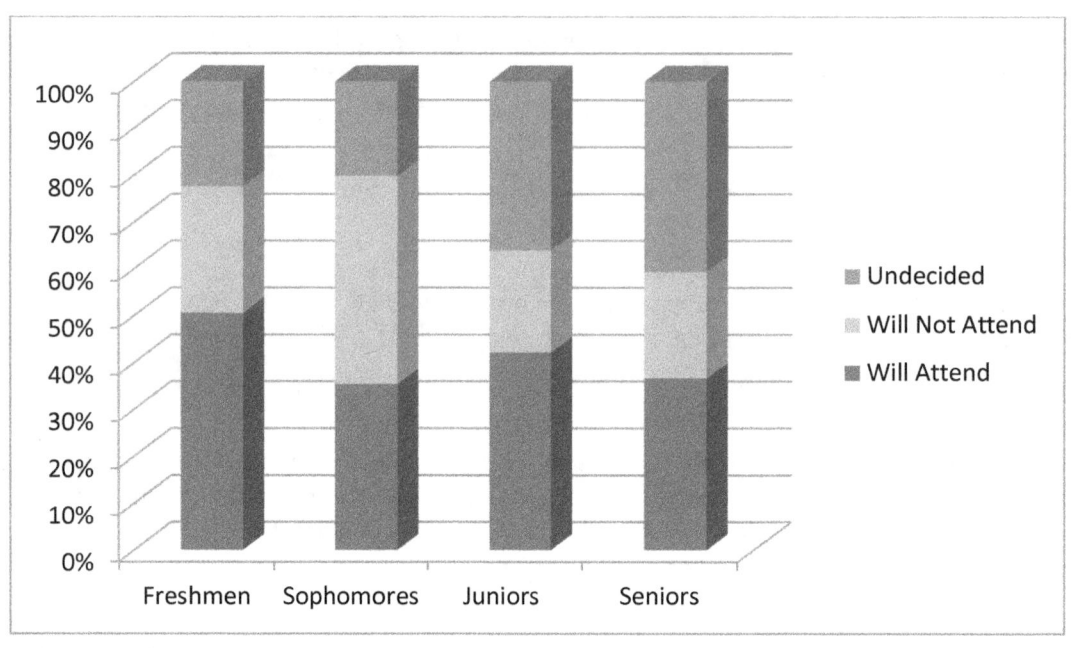

A) 25%   B) 35%   C) 45%   D) 55%

188) The pictograph below illustrates the results of a customer satisfaction survey by region. Each of the four regions has one salesperson. Salespeople in each region receive bonuses based on the amount of positive customer feedback they receive. If the salespeople from all four regions received $540 in bonuses in total, how much bonus money does the company pay each individual salesperson per satisfied customer?

Region 1	☺☺☺☺
Region 2	☺☺☺
Region 3	☺☺
Region 4	☺☺☺

Each ☺ represents positive feedback from 10 customers.

A) $4.00
B) $4.50
C) $4.90
D) $5.00

189) A dance judge awards a number from 1 to 10 to score dancers during a TV show. During one show, he judged five dancers and awarded the following scores: 9.9, 9.9, 8.2, 7.6 and 6.8. What was the median value of his scores for this show?

A) 8.2 B) 8.48 C) 9.9 D) 3.1

190) W = {2, 4, 8, 16, 32}
X = {4, 8, 12, 16, 20}
Y = {8, 16, 24, 32, 40}
Given sets W, X, and Y above, which of the following represents W ∩ X ∩ Y, the intersection of W and X with Y?
A) {2, 4, 8, 12, 16, 20, 24, 32, 40}
B) {4, 8, 16, 24, 32, 40}
C) {8, 16}
D) {8}

Algebra, Measurement, and Estimation

191) Expand the polynomial: $(x - 5)(3x + 8)$
A) $3x^2 - 7x - 40$ B) $3x^2 - 7x + 40$ C) $3x^2 + 23x - 40$ D) $3x^2 + 23x + 40$

192) If $\frac{3}{4}x - 2 = 4$, x = ?
A) $\frac{8}{3}$ B) $\frac{1}{8}$ C) 8 D) –8

193) Solve for x: $x^2 + 2x - 8 \leq 0$
A) $-4 \geq x \geq 2$
B) $-4 \geq x \leq 2$
C) $-4 \leq x \geq 2$
D) $-4 \leq x \leq 2$

194) If $x - 15 > 0$ and $y = x - 15$, then $y >$?
A) x B) $x + 15$ C) $x - 15$ D) 0

195) 2 inches on a scale drawing represents F feet. Which of the following equations represents $F + 1$ feet on the drawing?

A) $\frac{2(F+1)}{F}$ B) $\frac{(F+1)}{F}$ C) $\frac{2}{F+1}$ D) $\frac{2F}{F+1}$

196) The speed of sound in a recent experiment was 340,000 millimeters per second. How far did the sound travel in 1,000 seconds?
A) 3.4×10^5 millimeters
B) 3.4×10^6 millimeters
C) 3.4×10^7 millimeters
D) 3.4×10^8 millimeters

197) Which of the following is equivalent to the expression $2(x + 2)(x - 3)$ for all values of x?
A) $2x^2 - 2x - 12$ B) $2x^2 - 10x - 6$ C) $2x^2 + 2x - 12$ D) $2x^2 + 10x - 6$

198) Which of the following is a factor of: $2xy - 6x^2y + 4x^2y^2$
A) $(1 + 3x - 2xy)$ B) $(1 - 3x + 2xy)$ C) $(1 + 3x + 2xy)$ D) $(1 - 3x - 2xy)$

199) A construction company is building new homes on a housing development. It has an agreement with the municipality that H number of houses must be built every 30 days. If H number of houses are not built during the 30 day period, the company has to pay a penalty to the municipality of P dollars per house. The penalty is paid per house for the number of houses that fall short of the 30-day target. If A represents the actual number of houses built during the 30-day period, which equation below can be used to calculate the penalty for the 30-day period?
A) $(H - P) \times 30$
B) $(H - A) \times P$
C) $(A - H) \times 30$
D) $(A - H) \times P$

200) Perform the operation: $(5ab - 6a)(3ab^3 - 4b^2 - 3a)$
A) $15a^2b^4 - 20ab^3 - 15a^2b - 18a^2b^3 - 24ab^2 - 18a^2$
B) $15a^2b^4 - 20ab^3 - 15a^2b - 18a^2b^3 + 24ab^2 + 18a^2$
C) $15a^2b^4 - 20ab^3 - 15a^2b - 18a^2b^3 - 24ab^2 + 18a^2$
D) $15ab^4 - 20ab^3 - 15a^2b - 18a^2b^3 + 24ab^2 + 18a^2$

201) Which of the following is equivalent to $\frac{x}{5} \div \frac{9}{y}$?

A) $\frac{xy}{45}$
B) $\frac{9x}{5y}$
C) $\frac{1}{9} \times \frac{x}{5y}$
D) $\frac{1}{5} \times \frac{9}{5y}$

202) Which of the following values of x is a possible solution to the inequality?: $-3x + 14 < 5$
A) −3.1
B) 2.80
C) 2.25
D) 3.15

203) $(x - 2y)(2x^2 - y) = ?$
A) $2x^3 - 4x^2y + 2y^2 - xy$
B) $2x^3 + 2y^2 - 5xy$
C) $2x^3 - 4x^2y + 2y^2 + xy$
D) $2x^3 + 4x^2y + 2y^2 - xy$

204) What is the value of the expression $2x^2 + 5xy - y^2$ when x = 4 and y = −3?
A) −37
B) −19
C) 86
D) 101

205) If $6 + 8(2\sqrt{x} + 4) = 62$, then $\sqrt{x} = ?$
A) 3.25
B) 24
C) $\frac{3}{2}$
D) $\frac{2}{3}$

206) $\sqrt{18} \times \sqrt{8} = ?$
A) $18\sqrt{8}$
B) $\sqrt{26}$
C) $\sqrt{12}$
D) 12

207) If $2(3x - 1) = 4(x + 1) - 3$, what is the value of x?
A) ³/₂
B) 3
C) ²/₃
D) 2

208) The graph of a linear equation is shown below. Which one of the tables of values best represents the points on the graph?

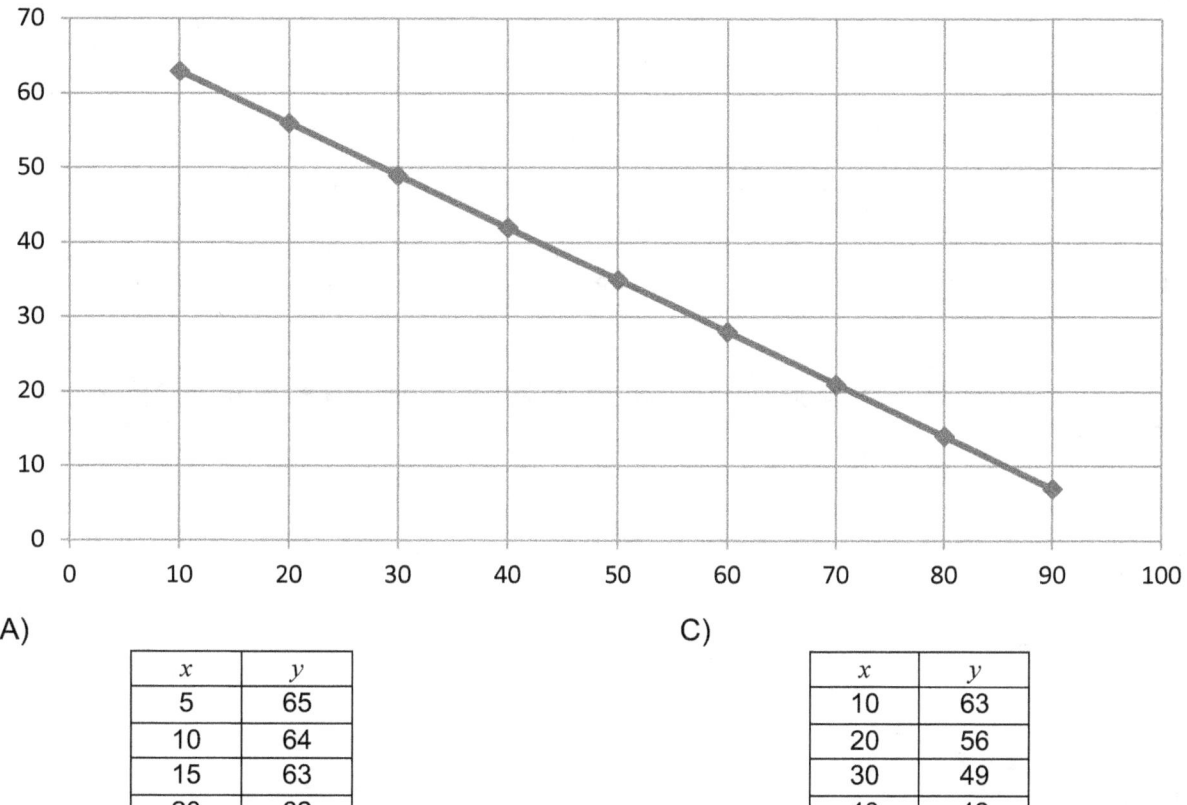

A)

x	y
5	65
10	64
15	63
20	62

C)

x	y
10	63
20	56
30	49
40	42

B)

x	y
5	68
15	60
25	52
35	54

D)

x	y
10	68
20	60
30	52
40	44

209) Triangles FGH and FGJ are right triangles. The lengths of FG, GH, and HJ are provided in units. What is the area of triangle FHJ in square units?

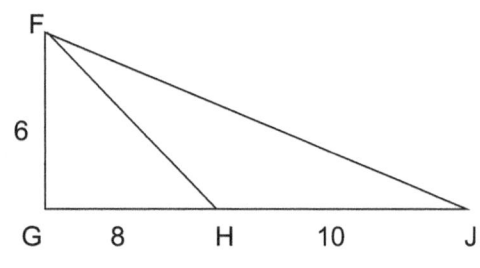

A) 24 B) 30 C) 48 D) 54

210) ABC is an isosceles triangle. Angle DAC is 109° and points A, B, and D are co-linear. What is the measurement of ∠C?

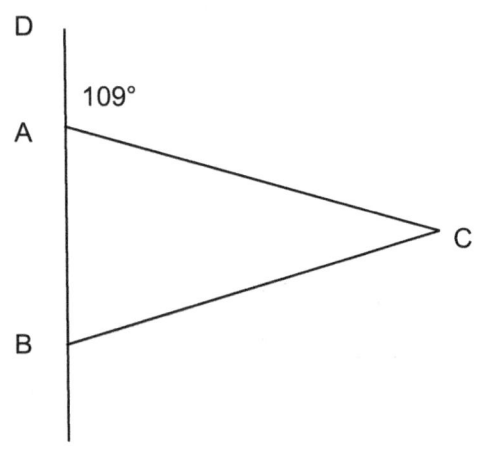

A) 71° B) 38° C) 138° D) 180°

211) If a circle has a diameter of 36, what is the area of the circle?
A) 18π B) 36π C) 72π D) 324π

212) $20 - \frac{3x}{4} \geq 17$, then $x \leq$?
A) −12 B) −4 C) −3 D) 4

213) A rectangular box has a base that is 5 inches wide and 6 inches long. The height of the box is 10 inches. What is the volume of the box?
A) 30 B) 110 C) 150 D) 300

214) Which of the following points lies on the graph of $10x + 3y = 29$?
A) (3, 2) B) (2, 3) C) (1, 6) D) (6, 1)

215) A vegetable grower wants to put to put wooden fence panels around the outside of her vegetable patch. Each panel is 1 yard in length. The patch is rectangular and is 12 yards long and 10 yards wide. How many panels are needed in order to enclose the vegetable patch?

A) 22 B) 44 C) 100 D) 120

216) Find the volume of a cone which has a radius of 3 and a height of 4.

A) 4π B) 12π C) $4\pi/3$ D) $3\pi/4$

217) Triangle ABC is a right-angled triangle, where side A and side B form the right angle, and side C is the hypotenuse. If A = 7 and C = 14, what is the length of side B?

A) 2 B) 7 C) $\sqrt{147}$ D) 147

218)

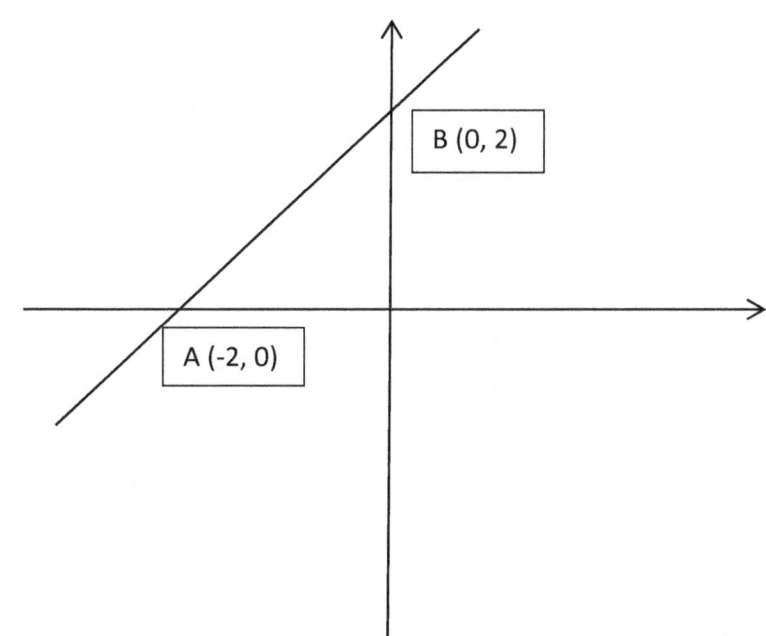

The line in the xy plane above is going to be shifted 5 units to the left and 4 units up. What are the coordinates of point B after the shift?

A) (–5, 6) B) (5, 6) C) (5, 4) D) (–7, 4)

219) A packaging company uses string to secure their packages prior to shipment. The string is tied around the entire length and entire width of the package, as shown in the following illustration:

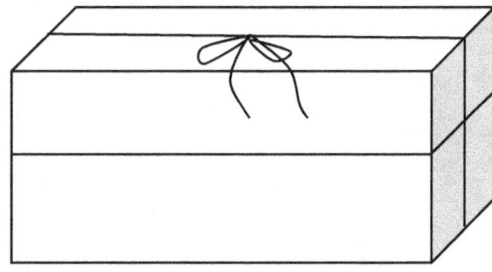

The box is ten inches in height, ten inches in depth, and twenty inches in length. An additional fifteen inches of string is needed to tie a bow on the top of the package. How much string is needed in total in order to tie up the entire package, including making the bow on the top?

A) 80　　　　　　　　B) 100　　　　　　　　C) 120　　　　　　　　D) 135

220) The triangle in the illustration below is an equilateral triangle. What is the measurement in degrees of angle *a*?

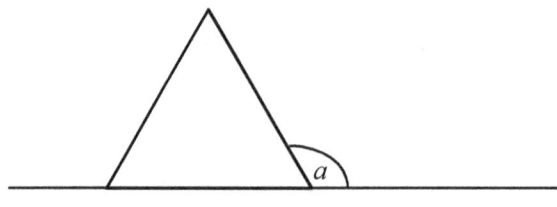

A) 45　　　　　　　　B) 60　　　　　　　　C) 120　　　　　　　　D) 180

221) The radius (R) of circle A is 5 centimeters. The radius of circle B is 3 centimeters. Which of the following statements is true?

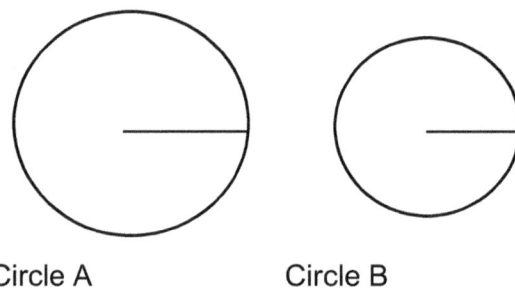

Circle A　　　　　　　　Circle B

A) The difference between the areas of the circles is 2.
B) The difference between the areas of the circles is 9π.
C) The difference between the circumferences of the circles is 2.
D) The difference between the circumferences of the circles is 4π.

222) In the standard (*x*, *y*) plane, what is the distance between $(4\sqrt{7}, -2)$ and $(7\sqrt{7}, 4)$?
A) $3\sqrt{11}$
B) 27
C) 36
D) 99

223) The graph below illustrates which of the following functions?

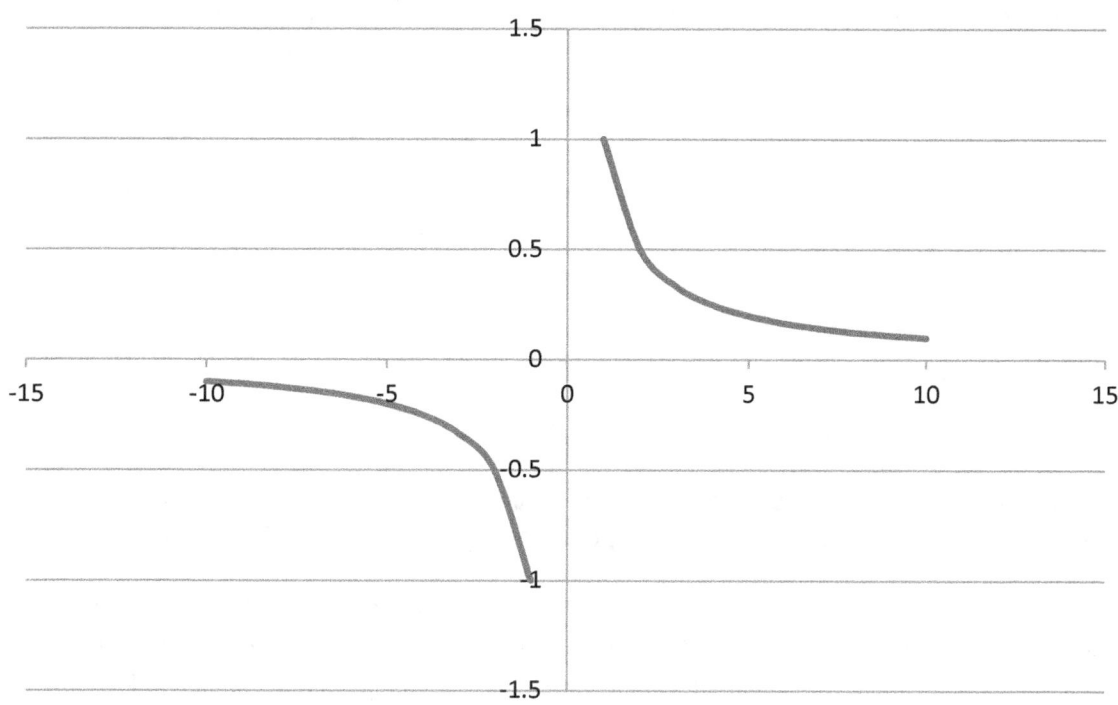

A) $f(x) = \frac{1}{x}$ B) $f(x) = \frac{x}{2}$ C) $f(x) = x^2 + 2$ D) $f(x) = x^2 - 2$

224) For the two functions $f_1(x)$ and $f_2(x)$, tables of vales are given below. What is the value of $f_2(f_1(2))$?

x	$f_1(x)$
1	3
2	5
3	7
4	9
5	11

x	$f_2(x)$
2	4
3	9
4	16
5	25
6	36

A) 4
B) 5
C) 9
D) 25

225) For the functions $f_2(x)$ listed below, x and y are integers greater than 1. If $f_1(x) = x^2$, which of the functions below has the greatest value for $f_1(f_2(x))$?

A) $f_2(x) = x/y$ B) $f_2(x) = y/x$ C) $f_2(x) = xy$ D) $f_2(x) = x - y$

226) The length of XZ in the illustration below s 10 units, sin 40° = 0.643, cos 40° = 0.776, and tan 40° = 0.839. Approximately how many units long is XY ?

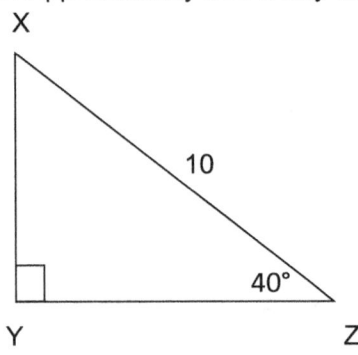

A) 0.643 B) 0.776 C) 7.76 D) 6.43

227) The cosine of a 41 degree angle is equal to which of the following?
A)   45° − 41° B)   the sine of 49° C) the tangent of 41° D) $(0.75470958)^2$

228) The line segment that runs from top of pole (P) to point (A) in the parking area forms a 70° angle. If the distance between the bottom of the pole (B) and point A in the parking area is 72 yards, what equation below calculates the distance in feet from the top of pole (P) to the bottom of the pole (B)?

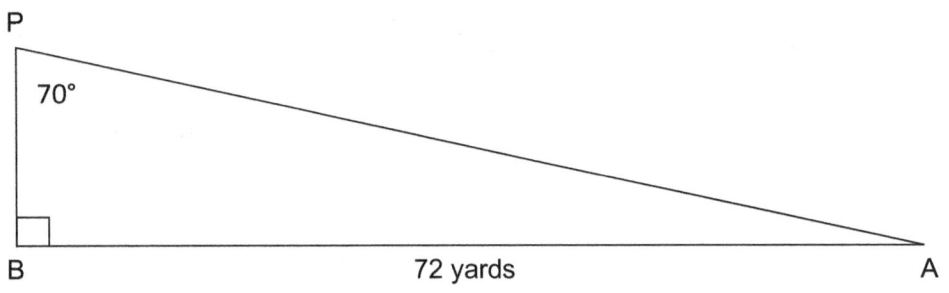

A) 72 × tan70° B) 72 ÷ tan70° C) 72 × cos70° D) 72 × sin70°

229) The cosine squared of angle A is 0.235040368. What is the sine squared of angle A?
A)   0.235040368 B) $\dfrac{1}{0.235040368}$ C) $(0.764959632)^2$ D) 0.764959632

230) In the figure below, the length of AC is 14 units, sin 55° = 0.8192, cos 55° = .5736, and tan 55° = 1.4281. Approximately how many units long is BC ?

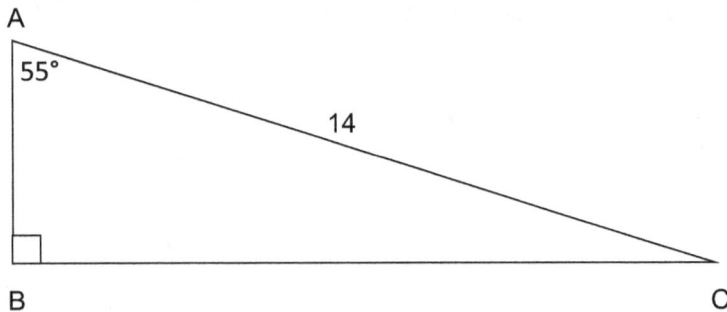

A) 0.8192 B) 19.9934 C) 8.0304 D) 11.469

231) Four members of a family are having a meal in a restaurant. They each have a main dish and a desert. The main dishes are all the same price each, and the deserts are also all the same price for each dessert. The main dishes cost $8 each. The total cost of their meal is $48. How much did each of their deserts cost?
A) $3.75 B) $4 C) $6 D) $22

232) If ⊖ is a special operation defined by (x ⊖ y) = (5x + 2y) and (6 ⊖ z) = 44, then z = ?
A) 32 B) 36 C) 7 D) 17

233) If $f(x) = x^2 + 3x - 8$, what is $f(x + 3)$?
A) $(x + 3)^2 + 3x - 8$
B) $(x + 3)^2 + 3(x + 3) - 8$
C) $x^2 + 3x - 5$
D) $3(x^2 + 3x - 8)$

234) $\sqrt{-16} = ?$
A) 4
B) –4
C) an imaginary number
D) Cannot be determined

235) A fuel tanker truck has a capacity of 1200 gallons. If it takes 75 minutes to fill the tanker truck, at what rate is it being filled?
A) 6.25 gallons per minute
B) 16 gallons per minute
C) 62.5 gallons per minute
D) 160 gallons per minute

236) Which of the following best describes the range of $y = -5x^3 - 40$?
A) All real numbers.
B) All real numbers except 0.
C) All real positive numbers.
D) All real negative numbers.

237) The term BPM, heartbeats per minute, measures how many heartbeats a person has every 60 seconds. To calculate BPM, the heartbeat is taken for ten seconds, represented by variable B. What equation is used to calculate BPM?
A) BPM ÷ 60 B) BPM ÷ 10 C) B6 D) B10

238) The ideal BPM of a healthy person is 60. What equation is used to calculate by how much a person's BPM exceeds the ideal BPM?
A) 60 + BPM B) 60 – BPM C) BPM + 60 D) BPM – 60

239) Shawn's final grade for a class is based on his grades from two projects, X and Y. Project X counts toward 45% of his final grade. Project Y counts toward 55% of his final grade. What equation can be used to calculate Shawn's final grade for this class?
A) .55X + .45Y B) .45X + .55Y C) (.45X + .55Y) ÷ 2 D) X + Y

240) The number of visitors a museum had on Tuesday (T) was twice as much as the number of visitors it had on Monday (M). The number of visitors it had on Wednesday (W) was 20% greater than that on Tuesday. Which equation can be used to calculate the total number of visitors for the three days?
A) M + 2T + W B) M + 1.2T + W C) W + .20W + 2T + M D) 5.4M

HiSET Math Practice Test Set 4 – Questions 241 to 320

Numerical Operations Problems

241) Yesterday a train traveled $117^{3}/_{4}$ miles. Today it traveled $102^{1}/_{6}$ miles. What is the difference between the distance traveled today and yesterday?
A) 15 miles
B) $15^{1}/_{4}$ miles
C) $15^{7}/_{12}$ miles
D) $15^{9}/_{12}$ miles

242) Sam is driving a truck at 70 miles per hour. At 10:30 am, he sees this sign:

Brownsville	35 miles
Dunnstun	70 miles
Farnam	140 miles
Georgetown	210 miles

After Sam sees the sign, he continues to drive at the same speed. At 11:00 am, how far will he be from Farnam?
A) He will be in Farnam.
B) He will be 35 miles from Farnam.
C) He will be 70 miles from Farnam.
D) He will be 105 miles from Farnam.

243) In a math class, $1/3$ of the students fail a test. If twelve students have failed the test, how many students are in the class in total?
A) 15
B) 16
C) 36
D) 38

244) Mark owns a bargain bookstore that sells every book for $5. Last week, his sales were $525. This week his sales figure was $600. How many more books did Mark sell this week, compared to last week?
A) 5
B) 15
C) 25
D) 75

245) Kieko needs to calculate 16% of 825. Which of the following formulas can she use?
A) 825 × 16
B) 160 × 825
C) 825 × 1600
D) 825 × 0.16

246) Wei Lei bought a shirt on sale. The original price of the shirt was $18, and he got a 40% discount. What was the sales price of the shirt?
A) $7.20
B) $10.80
C) $11.80
D) $17.60

247) Professor Smith uses a system of extra-credit points for his class. Extra-credit points can be offset against the points lost on an exam due to incorrect responses. David answered 18 questions incorrectly on the exam and lost 36 points. He then earned 25 extra credit points. By how much was his exam score ultimately lowered?
A) −11
B) 11
C) 18
D) 25

248) The county is proposing a 7.5% increase in its annual real estate tax. If the tax is currently $480 per year, how much would the tax be if the proposed increase is approved?
A) $444
B) $487
C) $516
D) $840

249) Mrs. Ramirez is inviting 12 children to her son's birthday party. The children will play pin the tail on the donkey. Mrs. Ramirez has already made 40 tails for the game. She wants to give each child 4 tails to play the game. How many more tails does she need to make?
A) 4　　　　　　　　B) 8　　　　　　　　C) 10　　　　　　　　D) 12

250) A class contains 20 students. On Tuesday 5% of the students were absent. On Wednesday 20% of the students were absent. How many more students were absent on Wednesday than on Tuesday?
A) 1　　　　　　　　B) 2　　　　　　　　C) 3　　　　　　　　D) 4

251) Records indicate that there were 12 hospitals in Johnson County in 1998, but this number had increased to 15 hospitals in 2016. There were 12 births per hospital in Johnson County in 1998. The total number of births in Johnson County was 240 in 2016. By what amount does the average number of births per hospital in Johnson County for 2016 exceed those for 1998?
A)　3 births per hospital　　　　　　　　B)　4 births per hospital
C)　15 births per hospital　　　　　　　　D)　16 births per hospital

252) Marta can walk one mile in 17 minutes. At this rate, how long would it take her to walk 5 miles?
A)　1 hour and 5 minutes　　　　　　　　B)　1 hour and 7 minutes
C)　1 hour and 8 minutes　　　　　　　　D)　1 hour and 25 minutes

253) Sam's final grade for a class is based on his scores from a midterm test (M), a project (P), and a final exam (F). The midterm test counts twice as much as the project, and the final exam counts twice as much as the midterm. Which mathematical expression below can be used to calculate Sam's final grade?
A) P + M + F　　　　B) P + M + 2F　　　　C) P + 2M + F　　　　D) P + 2M + 4F

254) Bart is riding his bike at a rate of 12 miles per hour. He arrives in the town of Wilmington at 3:00 pm. The town of Mount Pleasant is 50 miles from Wilmington. How far will Bart be from Mount Pleasant at 5:00 pm if he continues riding his bike at this speed?
A) 12 miles　　　　　B) 20 miles　　　　　C) 24 miles　　　　　D) 26 miles

255) A ticket office sold 360 more tickets on Friday than it did on Saturday. If the office sold 2570 tickets in total during Friday and Saturday, how many tickets did it sell on Friday?
A)　360　　　　　　　B) 1105　　　　　　　C) 1465　　　　　　　D) 1565

256) Tom's height increased by 10% this year. If Tom was 5 feet tall at the beginning of the year, how tall is he now?
A) 5 feet 1 inch　　　B) 5 feet 5 inches　　C) 5 feet 6 inches　　D) 5 feet 10 inches

257) Carlos buys 2 pairs of jeans for $22.98 each. He later decides to exchange both pairs of jeans for 3 sweaters which cost $15.50 each. Which equation can Carlos use to calculate the extra money he will have to pay for the exchange?
A) 2 × (22.98 - 15.50)　　　　　　　　B) 3 × (22.98 - 15.50)
C) (3 × 22.98) – (2 × 15.50)　　　　　　D) (3 × 15.50) – (2 × 22.98)

258) Jason does the high jump for his high school track and field team. His first jump is at 3.246 meters. His second is 3.331 meters, and his third is 3.328 meters. If the height of each jump is rounded up or down to the nearest one-hundredth of a meter (also called a centimeter), what is the estimate of the total height for all three jumps combined?
A) 9.80   B) 9.89   C) 9.90   D) 9.91

259) Use the table below to answer the question that follows.

Regional Railway Train Service	
Departure Time	Arrival Time
9:50 am	10:36 am
11:15 am	12:01 pm
12:30 pm	1:16 pm
2:15 pm	3:01 pm
?	5:51 pm

The journey on the Regional Railway is always exactly the same duration. What is the missing time in the chart above?
A) 3:30 pm   B) 4:15 pm   C) 4:30 pm   D) 5:05 pm

260) Captain Smith needs to purchase rope for his fleet of yachts. He owns 26 yachts and needs 6 feet 10 inches of rope for each one. How much rope does he need in total?
A) 152 feet   B) 177 feet 8 inches   C) 257 feet 8 inches   D) 260 feet

Data Interpretation and Probability Problems

261) Find the value of x that solves the following proportion: $3/6 = x/14$
A) 3   B) 6   C) 7   D) 8

262) In a shipment of 100 mp3 players, 1% are faulty. What is the ratio of non-faulty mp3 players to faulty mp3 players?
A) 1:100   B) 100:1   C) 99:100   D) 99:1

263) Retail prices for a particular item at four different stores are provided below. Calculate the variance in the retail prices: $12; $14; $10; $8
A) $5   B) $6   C) $9   D) $2.24

264) A bag contains 5 red balloons, 10 orange balloons, 8 green balloons, and 12 purple balloons. If a balloon is drawn from the bag at random, what is the probability that it will be orange?

A) $\frac{2}{7}$ B) $\frac{1}{4}$ C) $\frac{1}{10}$ D) $\frac{1}{35}$

265) A deck of cards contains 13 hearts, 13 diamonds, 13 clubs, and 13 spades. Cards are selected from the deck at random. Once selected, the cards are discarded and are not placed back into the deck. Two spades, one heart, and a club are drawn from the deck. What is the probability that the next card drawn from the deck will be a heart?

A) $1/13$ B) $1/12$ C) $13/52$ D) $1/4$

266) The chart below shows data on the number of vehicles involved in accidents in Cedar Valley. Pick-ups and vans were involved in approximately what percentage of total vehicle accidents on June 1?

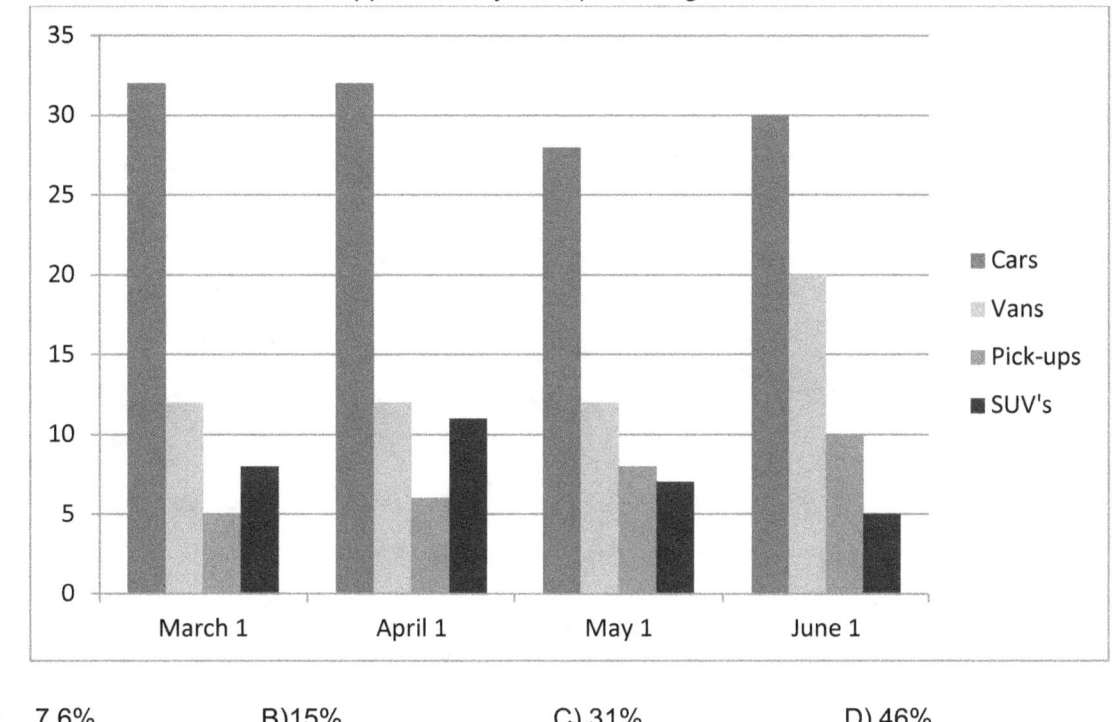

A) 7.6%   B) 15%   C) 31%   D) 46%

267) The pictograph below shows the number of traffic violations that occur every week in a certain city. The fine for speeding violations is $50 per violation. The fine for other violations is $20 per violation. The total collected for all three types of violations was $6,000. What is the fine for each parking violation?

Speeding	☆ ☆
Parking	☆
Other violations	☆ ☆ ☆

Each ☆ represents 30 violations.

A) $20   B) $30   C) $40   D) $100

268) The graph below shows the relationship between the number of days of rain per month and the amount of people who exercise outdoors per month. What relationship can be observed?

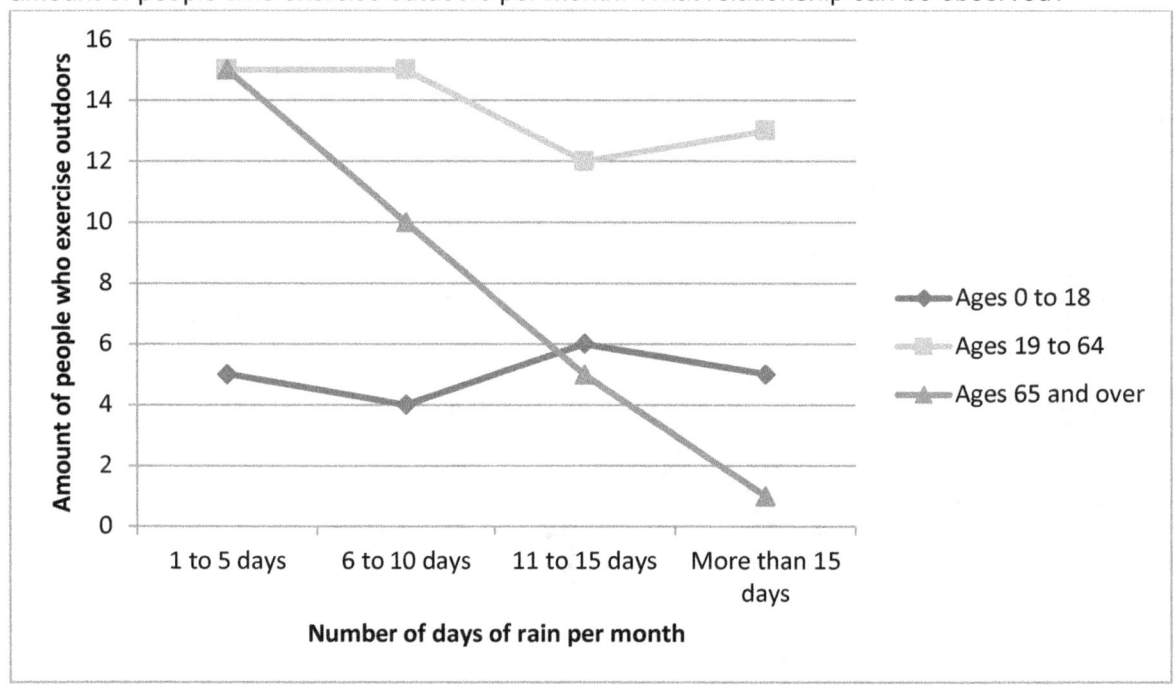

A) Young children are reliant upon an adult in order to exercise outdoors.
B) The exercise habits of working age people seem to fluctuate proportionately to the amount of rainfall.
C) In the 19 to 64 age group, there is a negative relationship between the number of days of rain and the amount of people who exercise outdoors.
D) People aged 65 and over seem less inclined to exercise outdoors when there is more rain.

269) Seven members of a support group are trying to gain weight. So far, the weight gain in kilograms for each of the seven members of the group is: 12, 15, 3, 7, 21, 14, and 12. What is the range of the amount of weight gain for this support group?

A) 18   B) 12   C) 14   D) 7

270) Looking at our seven group members from question 269 above, what is the mode?

A) 18   B) 12   C) 14   D) 7

Algebra, Measurement, and Estimation

271) Solve for x: $x^2 + 4x + 3 > 0$
A) $x < -3$ or $x > -1$
B) $x < -3$ or $x < -1$
C) $x > -3$ or $x < -1$
D) $x > -3$ or $x > -1$

272) $(x - 9y)^2 = ?$
A) $x^2 + 81y^2$
B) $x^2 - 18xy - 18y^2$
C) $x^2 - 18xy + 81y^2$
D) $x^2 + 18xy - 81y^2$

273) $6 + \frac{x}{4} \geq 22$, then $x \geq$?
A) −8
B) 64
C) −64
D) 128

274) $(x^2 - x - 12) \div (x - 4) = ?$
A) $(x + 3)$
B) $(x - 3)$
C) $(-x + 3)$
D) $(-x - 3)$

275) Which of the following expressions is equivalent to: $18xy - 24x^2y - 48y^2x^2$?
A) $6xy(3 - 4x - 8xy)$
B) $3xy(6 - 8x - 16xy)$
C) $6x^2y(3 - 4 - 8y)$
D) $6xy(3 - 4x + 8xy)$

276) $\sqrt{15x^3} \times \sqrt{8x^2}$
A) $\sqrt{23x^5}$
B) $2x^2\sqrt{30x^3}$
C) $2x^2\sqrt{30x}$
D) $\sqrt{23x^6}$

277) Which one of the following is a solution to the following ordered pairs of equations?
$-3x - 1 = y$
$x + 7 = y$
A) (5, −2)
B) (−2, 5)
C) (2, 5)
D) (5, 2)

278) $x^{-4} = ?$
A) $4\sqrt{x}$
B) $\sqrt[-4]{x}$
C) $x^4 \div 1$
D) $1 \div x^4$

279) If $5(4\sqrt{x} - 8) = 40$, then $x = ?$
A) $\frac{5}{12}$
B) 4
C) 16
D) $\sqrt{\frac{5}{12}}$

280) $\sqrt[3]{\frac{8}{27}} = ?$
A) $\frac{2}{3}$
B) $\frac{4}{9}$
C) $\frac{2}{9}$
D) $\frac{\sqrt{8}}{9}$

281) For all $a \neq b$, $\frac{5a/b}{2a/(a-b)} = ?$

A) $\frac{10a^2}{ab - b^2}$

B) $\frac{a - b}{2b}$

C) $\frac{5a - 5b}{2}$

D) $\frac{5a - 5b}{2b}$

282) The line on the xy-graph below forms the diameter of the circle. What is the area of the circle?

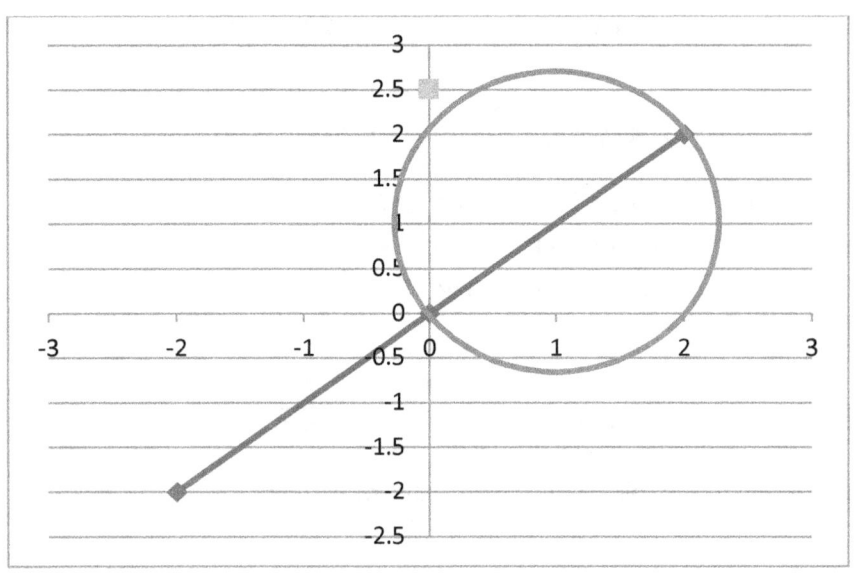

A) π B) 2π C) $\frac{\pi}{2}$ D) 2.5π

283) Which one of the scatterplots below most strongly suggests a negative linear relationship between x and y?

A)

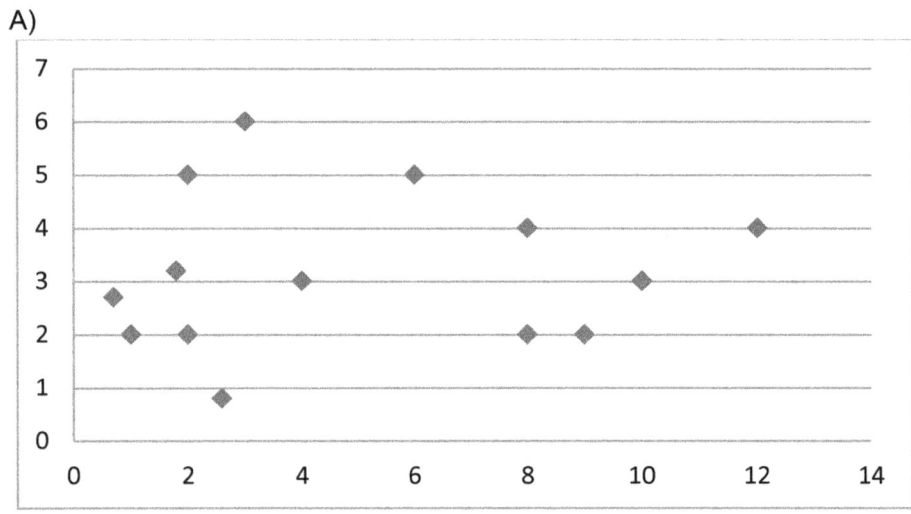

B)

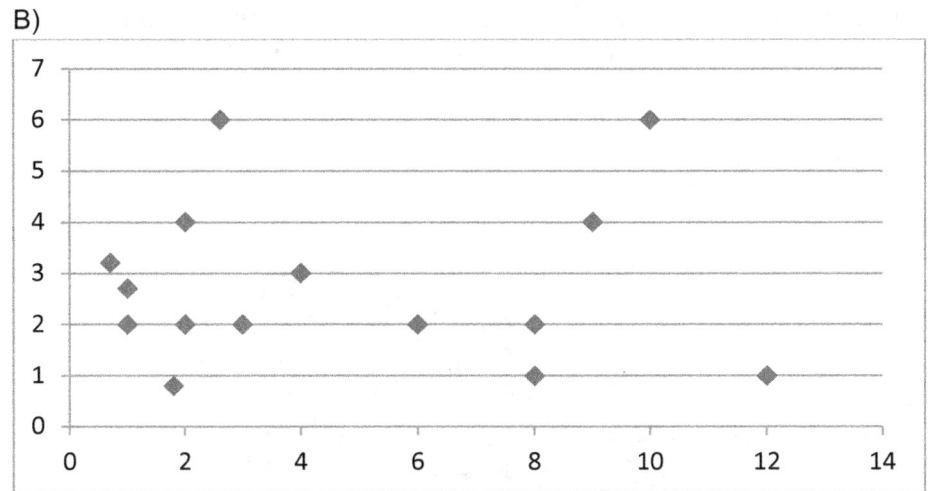

C)

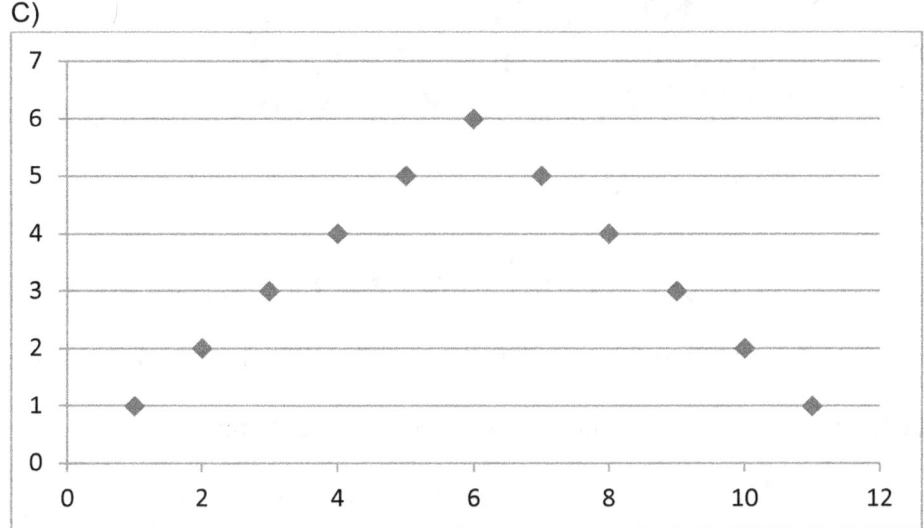

D)

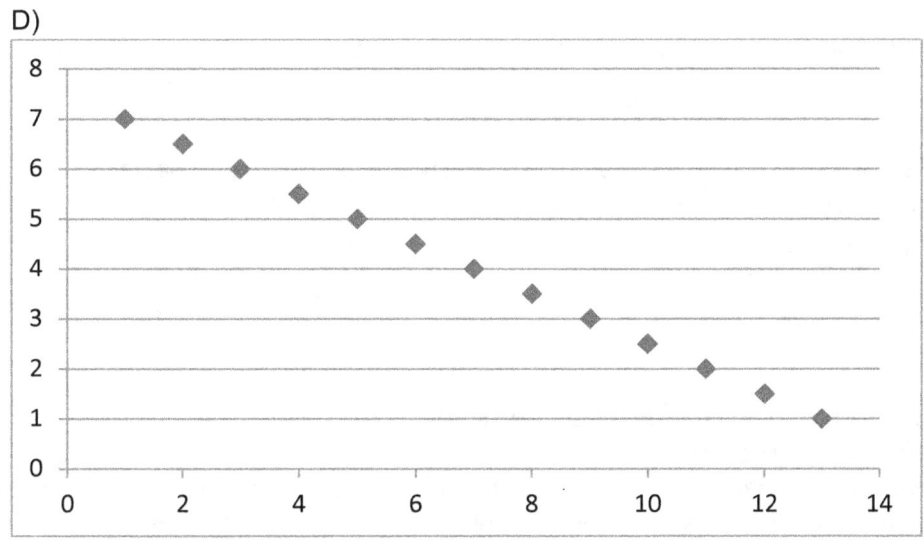

284) An airplane flew at a constant speed, traveling 780 miles in 2 hours. The graph below shows the total miles the airplane traveled in 20 minute intervals. What is the domain and range of this relation? Is this relation a function?

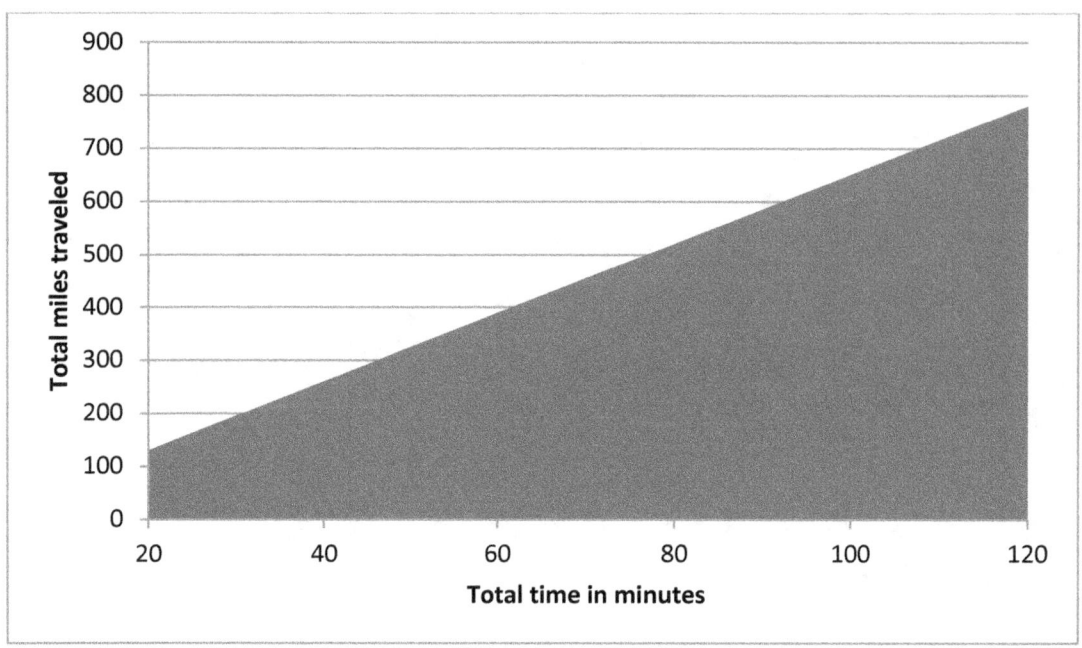

A) The domain is {20, 40, 60, 100, 120}; the range is {130, 260, 390, 520, 650, 780}; the relation is not a function.
B) The domain is {130, 260, 390, 520, 650, 780}; the range is {20, 40, 60, 100, 120}; the relation is not a function.
C) The domain is {130, 260, 390, 520, 650, 780}; the range is {20, 40, 60, 100, 120}; the relation is a function.
D) The domain is {20, 40, 60, 100, 120}; the range is {130, 260, 390, 520, 650, 780}; the relation is a function.

285) Find the area of the right triangle that has a base of 4 and a height of 15.

A) 20 B) 30 C) 60 D) 120

286) A small circle has a radius of 5 inches, and a larger circle has a radius of 8 inches. What is the difference in inches between the circumferences of the two circles?
A) 3
B) 6
C) 6π
D) 9π

287) Which of the following statements about isosceles triangles is true?
A) Isosceles triangles have two equal sides.
B) When an altitude is drawn in an isosceles triangle, two equilateral triangles are formed.
C) The base of an isosceles triangle must be shorter than the length of each of the other two sides.
D) The sum of the measurements of the interior angles of an isosceles triangle must be equal to 360°.

288) The illustration below shows a right circular cone. The entire cone has a base radius of 9 and a height of 18.

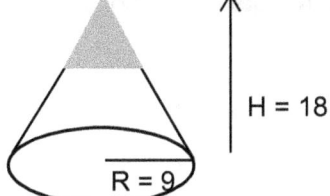

The shaded portion at the top of the cone has a height of 6. What is the volume of the shaded portion?

A) 18π
B) 36π
C) 48π
D) 486π

289) The diagram below depicts a cell phone tower. The height of the tower from point B at the center of its base to point T at the top is 30 meters, and the distance from point B of the tower to point A on the ground is 18 meters. What is the approximate distance from point A on the ground to the top (T) of the cell phone tower?

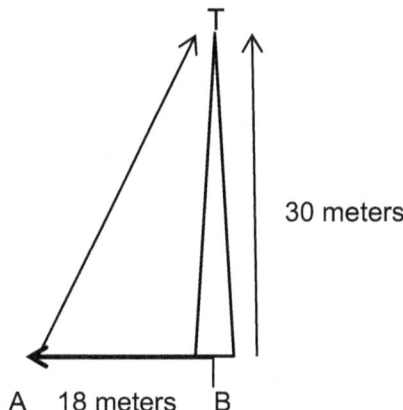

A) 10 meters
B) 20 meters
C) 30 meters
D) 35 meters

290) The graph of a line is shown on the xy plane below. The point that has the x-coordinate of 160 is not shown. What is the corresponding y-coordinate of that point?

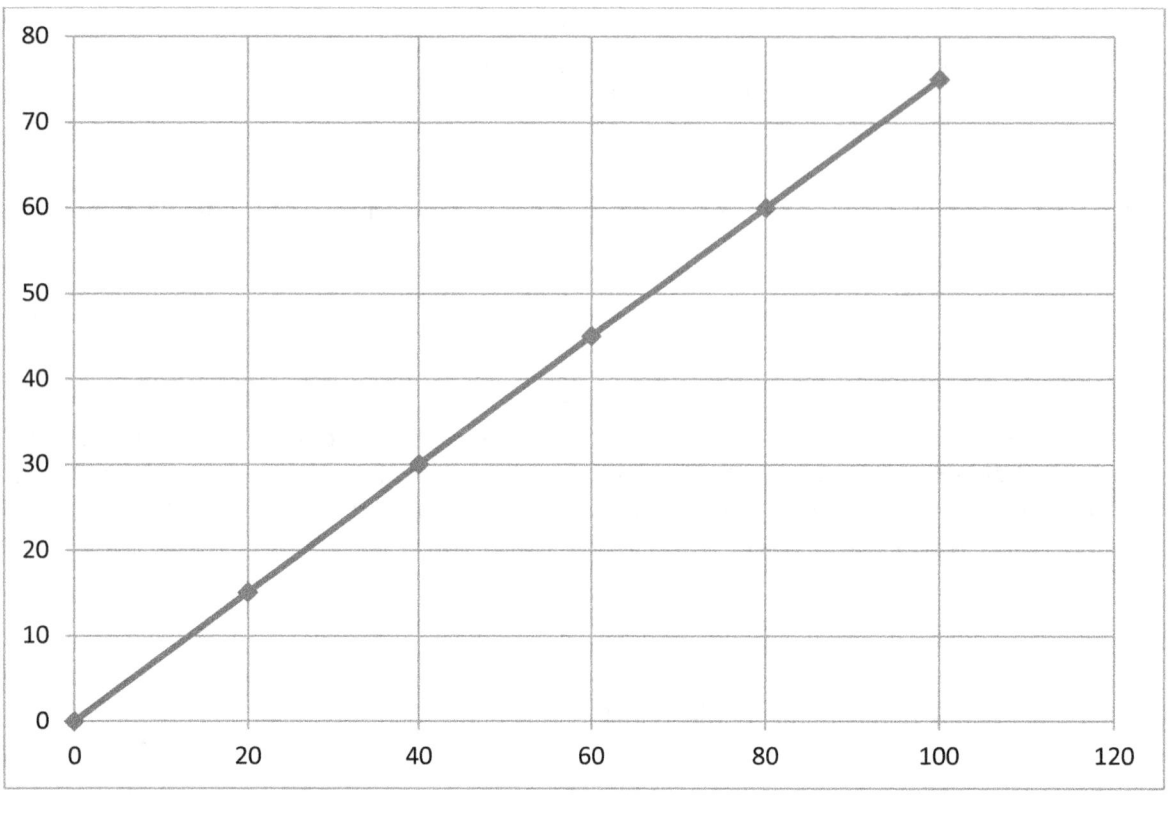

A) 115 B) 120 C) 125 D) 130

291) $-|5-8| = ?$
A) −13 B) 13 C) −3 D) 3

292) A company has decided to remodel their offices. They currently have 3 offices that measure 10 feet by 10 feet each and a common area that also measures 10 feet by 10 feet. When the offices are remodeled, there will be one large office that will be 20 feet by 10 feet and two small offices that will each be 10 feet by 8 feet. The remaining space is to be allocated to the new common area. What are the dimensions of the new common area?

A) 4×10 B) 8×10 C) 10×10 D) 4×8

293) Triangle QRS is a right-angled triangle. Side Q and side R form the right angle, and side S is the hypotenuse. If Q = 3 and R = 2, what is the length of side S?

A) 5 B) $\sqrt{5}$ C) $\sqrt{13}$ D) 13

294) Liz wants to put new vinyl flooring in her kitchen. She will buy the flooring in square pieces that measure 1 square foot each. The entire room is 8 feet by 12 feet. The cupboards are two feet deep from front to back. Flooring will not be put under the cupboards. A diagram of her kitchen is provided.

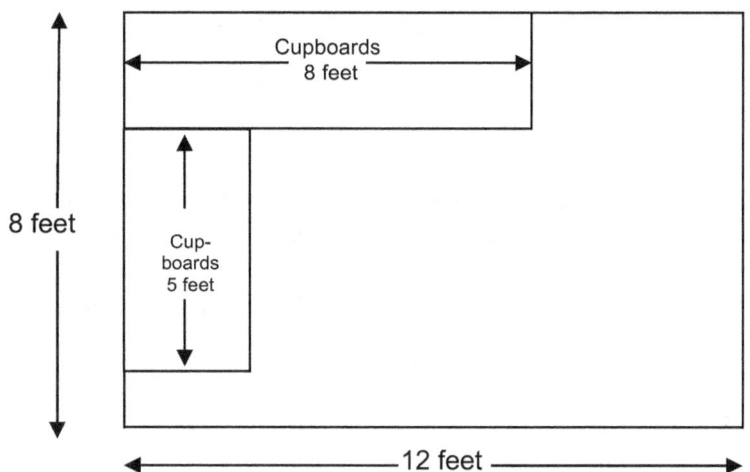

How many pieces of vinyl will Liz need to cover her floor?

A) 120 B) 96 C) 70 D) 84

295) The diagram below shows a figure made from a semicircle, a rectangle, and an equilateral triangle. The rectangle has a length of 18 inches and a width of 10 inches. What is the perimeter of the figure?

A) 56 inches + 5π inches
B) 56 inches + 10π inches
C) 56 inches + 12.5π inches
D) 56 inches + 25π inches

296) The illustration below shows a pyramid with a base width of 3, a base length of 5, and a volume of 30. What is the height of the pyramid?

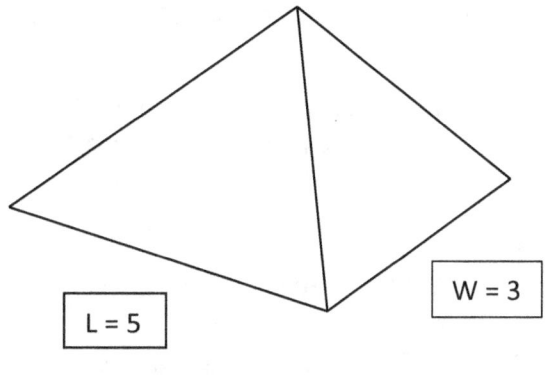

A) 2 B) 3 C) 5 D) 6

297) Find the x and y intercepts of the following equation: $5x^2 + 4y^2 = 120$

A) $(0, \sqrt{30})$ and $(\sqrt{24}, 0)$ B) $(0, 30)$ and $(24, 0)$

C) $(\sqrt{24}, 0)$ and $(0, \sqrt{30})$ D) $(30, 0)$ and $(0, 24)$

298) Consider a two-dimensional linear graph where x = 4 and y = 15. The line crosses the y axis at 3. What is the slope of this line?

A) $\frac{1}{15}$ B) 3   C) $-\frac{1}{3}$ D) –3

299) The perimeter of a rectangle is 48 meters. If the width were doubled and the length were increased by 5 meters, the perimeter would be 92 meters. What are the length (L) and width (W) of the original rectangle?

A) W = 17, L = 7   B) W = 7, L = 17   C) W = 34, L = 14   D) W = 24, L = 46

300) $4 = \log_4 256$ is equivalent to which of the following?

A) 4^4 B) 8   C) $4\sqrt{256}$ D) $\frac{256}{4}$

301) The graph below illustrates which of the following functions?

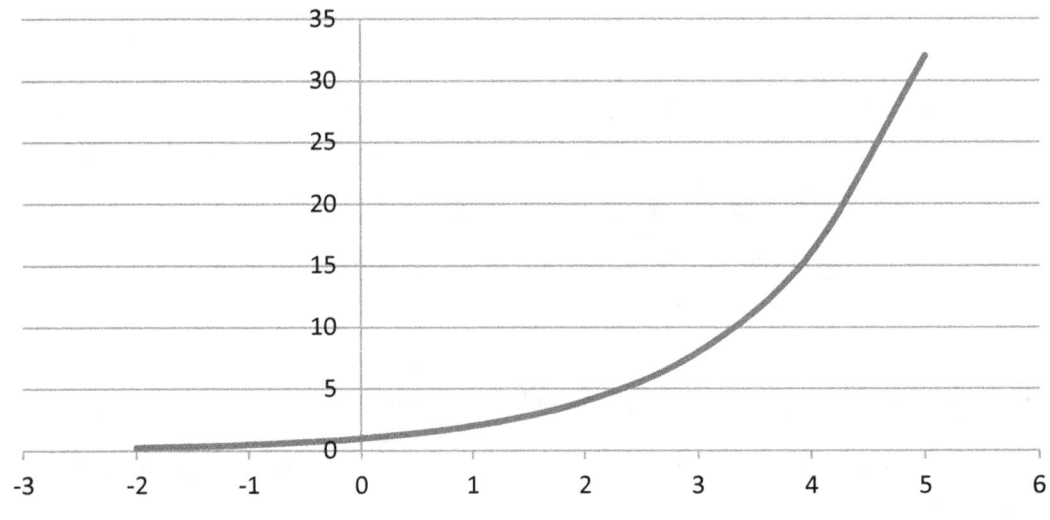

A) $f(x) = \frac{1}{x}$ B) $f(x) = 2^x$ C) $f(x) = \sqrt{x}$ D) $f(x) = x^2$

302) Express the following as a logarithmic function and solve for x: $4^{2x} = 64$

A) $2x = \log_4 64$; x = 3   B) $2x = \log_4 64$; x = 1.5

C) $4 = \log_{2x} 64$; x = 3   D) $4 = \log_{2x} 64$; x = 1.5

303) If $f_1(x) = x^2 + x$, what is the value of $f_1(5)$?

A) 5   B) 10   C) 25   D) 30

304) The sine squared of angle B is 0.1403301. What is the cosine squared of angle B?

A) 0.8596699   B) $(0.8596699)^2$ C) 0.1403301   D) 1 + 0.1403301

305) If sin A = 0.1908, then $\cos^2 A$ = ?
A) 0.036405
B) 0.8092
C) 0.0963595
D) 0.963595

306) The street from the court house (C) to the park (P) forms a 65° angle. The park (P) is 4 miles from the court house (C). What equation calculates the distance from the gas station (G) to the park?

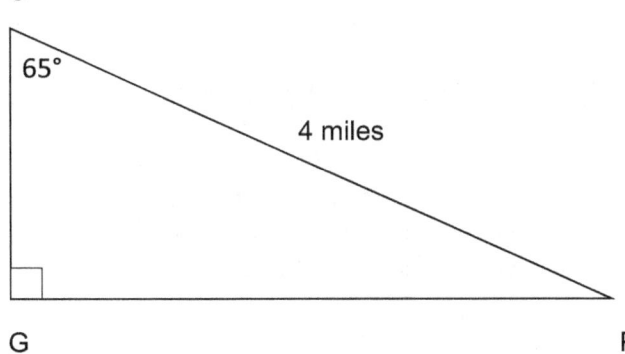

A) 4 × tan65°
B) 4 ÷ tan65°
C) 4 × cos65°
D) 4 × sin65°

307) The radius of a circle is 16 and the radians of the subtended angle measure $3\pi/4$. What is the length of the arc subtending the central angle?
A) 3π
B) 4π
C) 12π
D) 16π

308) Which of the following points lies on the graph of $5x + 6y = 34$?
A) (4, 2)
B) (1, 5)
C) (2, 4)
D) (−3, 8)

309) In the xy plane, line L passes through (0, 0) is perpendicular to line K. Line K is represented by the following equation: $y = 5x + 0$. The equation of line L could be which one of the following?
A) $y = -5x$
B) $y = \frac{1}{5}x + 0$
C) $y = 5x + 0$
D) $y = -\frac{1}{5}x$

310) The price of widgets is $2 each and the price of whatsits is $25 each. Zafira bought widgets and whatsits in one transaction, and she paid $85 in total. If she bought 3 whatsits, how many widgets did she buy?
A) 2
B) 3
C) 5
D) 8

311) If $f(x) = x \div (1 + x)$ and $g(x) = 1 \div x$, what is the domain of the function $f + g$?
A) {−1, 0}
B) All real numbers.
C) All real numbers except 0.
D) All real numbers except 0 and −1.

312) Consider the quadratic function $(x) = ax^2 + bx + c$. In this problem, $f(x) = y$. If m is a constant, how many solutions exist for the equation $ax^2 + bx + c = mx$?
A) 0
B) 1
C) 2
D) Cannot be determined

313) $\sqrt{3}$ is equivalent to what number in exponential notation?
A) $3^{\frac{1}{4}}$
B) $3^{\frac{1}{2}}$
C) $1^{\sqrt{3}}$
D) 3^0

314) A participant in a 100 mile endurance event ran at a speed of 5 miles per hour for the first 80 miles of the event and *x* miles per hour for the last 20 miles of the event. What equation represents the participant's average speed for the entire event?
A)   100 ÷ [(80 ÷ 5) + (20 ÷ *x*)]
B)   100 × [(80 ÷ 5) + (20 ÷ *x*)]
C)   100 ÷ [(80 × 5) + (20 × *x*)]
D)   100 × [(80 × 5) + (20 × *x*)]

315) Davina is using a recipe that requires 2 cups of sugar for every one-third cup of butter. If she uses 12 cups of sugar, how many cups of butter should she use?
A) 6 cups of butter   B) 3 cups of butter   C) 2 cups of butter   D) ⅓ cup of butter

316) A driver travels at 60 miles per hour for two and a half hours before her car fails to start at a service station. She has to wait two hours while the car is repaired before she can continue driving. She then drives at 75 miles an hour for the remainder of her journey. She is traveling to Denver, and her journey is 240 miles in total. If she left home at 6:00 am, what time will she arrive in Denver?

A)   9:30 am   B) 11:30 am   C) 11:42 am   D) 11:50 am

317) A clothing store sells jackets and jeans at a discount during a sales period. T represents the number of jackets sold and N represents the number of jeans sold. The total amount of money the store collected for sales of jeans and jackets during the sales period was $4,000. The amount of money earned from selling jackets was one-third of that earned from selling jeans. The jeans sold for $20 a pair. How many pairs of jeans did the store sell during the sales period?
A) 15   B)   20   C)   150   D) 200

318) Which of the following steps will solve the equation for *x*: 18 = 3(*x* + 5)
A)   Subtract 5 from each side of the equation, and then divide both sides by 3.
B)   Subtract 18 from each side of the equation, and then divide both sides by 5.
C)   Multiply both *x* and 5 by 3 on the right side of the equation. Then subtract 15 from each side of the equation.
D)   Divide each side of the equation by 3. Then subtract 5 from both sides of the equation.

319) The number of bottles of soda that a soft drink factory can produce during *D* number of days using production method A is represented by the equation: D^5 + 12,000. Alternatively, the number of bottles of soda that can be produced using method B is represented by this equation: *D* × 10,000. What is the largest number of bottles of soda that can be produced by the factory during a 10 day period?
A)   10,000   B) 12,000   C) 100,000   D) 112,000

320) Milk fills a tank at a diary at a rate of 3.5 gallons per minute. If the tank has a 500 gallon capacity, approximately how long will it take to fill the tank?
A) 2 hours and 23 minutes
B) 2 hours and 13 minutes
C) 1 hour and 43 minutes
D) 1 hour and 42 minutes

HiSET Math Practice Test Set 5 – Questions 321 to 400

Numerical Operations Problems

321) Use the diagram below to answer the question that follows.

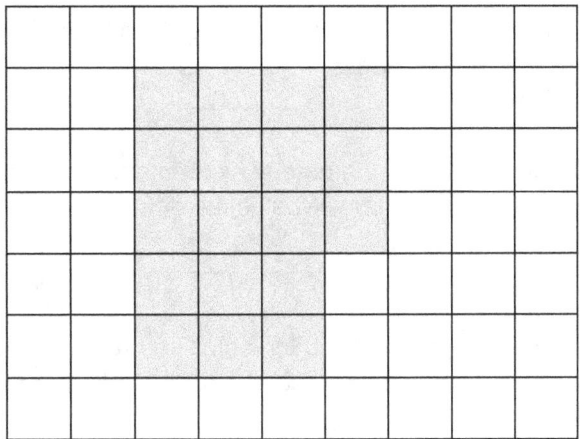

Each square in the diagram above is one foot wide and one foot long. The gray area of the diagram represents the layout of New Town's water reservoir. What is the perimeter in feet of the reservoir?
A) 16 feet B) 17 feet C) 18 feet D) 20 feet

322) Use the table below to answer the question that follows.

Part	Total Number of Questions	Number of Questions Answered Correctly
1	15	12
2	25	20
3	35	32
4	45	32

Chantelle took a test that had four parts. The total number of questions on each part is given in the table above, as is the number of questions Chantelle answered correctly. What was Chantelle's percentage of correct answers for the entire test?
A) 75% B) 80% C) 86% D) 90%

323) It takes Martha 4 hours and 10 minutes to knit one woolen cap. At this rate, how long will it take her to knit 12 caps?
A) 40 hours B) 42 hours C) 46 hours D) 50 hours

324) The Jones family needs to dig a new well. The well will be 525 feet deep, and it will be topped with a windmill which will be 95 feet in height. What is the distance from the deepest point of the well to the top of the windmill?
A) 95 feet B) 430 feet C) 525 feet D) 620 feet

325) Mrs. Thompson is having a birthday party for her son. She is going to give balloons to the children. She has one bag that contains 13 balloons, another that contains 22 balloons, and a third that contains 25 balloons. If 12 children are going to attend the party including her son, and the total amount of balloons is to be divided equally among all of the children, how many balloons will each child receive?
A) 3 B) 4 C) 5 D) 6

326) A bookstore is offering a 15% discount on books. Janet's purchase would be $90 at the normal price. How much will she pay after the discount?
A) $75.50 B) $76.50 C) $77.50 D) $85.50

327) John is measuring plant growth as part of a botany experiment. Last week, his plant grew 7¾ inches, but this week his plant grew 10½ inches. What is the difference in growth in inches between the two weeks?
A) 2¼ inches B) 2½ inches C) 2¾ inches D) 3¼ inches

328) Patty works 23 hours a week at a part time job for which she receives $7.50 an hour. She then gets a raise, after which she earns $184 per week. She continues to work 23 hours per week. How much did her hourly pay increase?
A) 50 cents an hour B) 75 cents an hour C) $1.00 an hour D) $8.00 an hour

329) Sheng Li is driving at 70 miles per hour. At 10:00 am, he sees this sign:

Washington	140 miles
Yorkville	105 miles
Zorster	210 miles

He continues driving at the same speed. Where will Sheng Li be at 11:00 am?
A) 70 miles from Washington
B) 105 miles from Washington
C) 75 miles from Yorkville
D) 80 miles from Yorkville

330) Mayumi spent the day counting cars for her job as a traffic controller. In the morning she counted 114 more cars than she did in the afternoon. If she counted 300 cars in total that day, how many cars did she count in the morning?
A) 90
B) 93
C) 114
D) 207

331) Tiffany buys five pairs of socks for $2.50 each. The next day, she decides to exchange these five pairs of socks for four different pairs that cost $3 each. She uses this equation to calculate her refund: (5 × $2.50) – (4 × $3). Which equation below could she have used instead?
A) (5 × 4) – (3 × 2.50)
B) $2.50 – 4($3 - $2.50)
C) (5 × 4) + (3 × 2.50)
D) $3 – (4 × $2.50)

332) Mr. Carlson needs to calculate 35% of 90. To do so, he uses the following equation: $\frac{35 \times 90}{100}$

Which of the following could he also have used?
A) (35 × 90) ÷ 100
B) (35 ÷ 90) × 100
C) (35 − 90) × 100
D) 90 × .0035

333) Use the table below to answer the question that follows.

Waterloo Station Bus Timetable	
Departure Time	Arrival Time
9:18 am	11:06 am
10:32 am	12:20 pm
11:52 am	?
1:03 pm	2:51 pm

The bus journeys from Waterloo Station to a nearby town are always the same duration. What time is missing from the above timetable?
A) 12:40 pm   B) 1:34 pm   C) 1:40 pm   D) 1:48 pm

334) Item C costs 20% more per pound than item B. If a 12 pound container of item B costs $48, what is the cost per pound of item C?
A) $4.12   B) $4.20   C) $4.60   D) $4.80

335) The cost of a photography course is $20 per week plus a $5 fee per week for review of photographs and administration. What is the total cost of the course and fees for W weeks?
A) $20W   B) $25W   C) $20 + 5W   D) $5 + 20W

336) Which of the following shows the numbers ordered from least to greatest?
A) $-1/4$, $1/8$, $1/6$, 1
B) $-1/4$, $1/8$, 1, $1/6$
C) $-1/4$, $1/6$, $1/8$, 1
D) $-1/4$, 1, $1/8$, $1/6$

337) The graph below shows the relationship between the total number of hamburgers a restaurant sells and the total sales in dollars for the hamburgers. The cost of shakes, where c is the cost and s is the number of shakes, is represented by the following equation: $c = \frac{9}{4}s$. Which of the following best estimates difference between the cost of one hamburger and the cost of one shake?

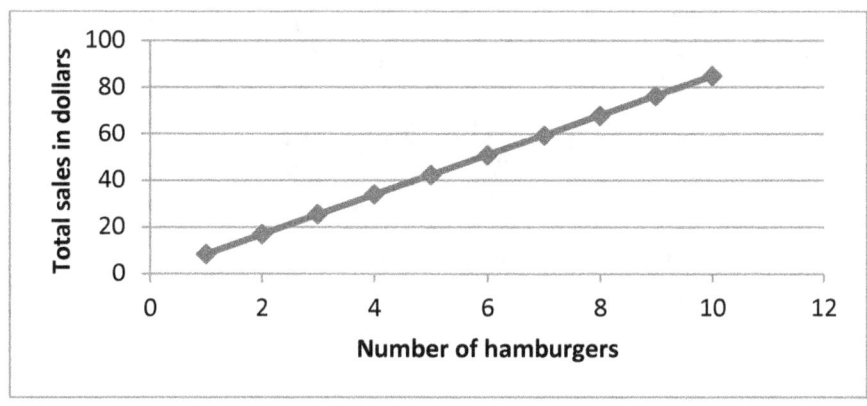

A) $2.25   B) $4.25   C) $6.25   D) $8.20

338) What is 12.86749 rounded to the nearest tenth?
A) 10   B) 12   C) 12.8   D) 12.9

339) Point B is at 0.35 on a number line and the distance between point B and point C is 1.2. Which of the following could be the location of point C?
A) 0.23 B) 0.47 C) –0.85 D) 0.85

340) What is the sum of 15.845 + 8.21 to the nearest integer?
A) 24.055 B) 24 C) 23 D) 22

Data Interpretation and Probability Problems

341) At an elementary school, 3 out of ten students are taking an art class. If the school has 650 students in total, how many total students are taking an art class?
A) 65 B) 130 C) 195 D) 217

342) Find the median of the following data set: 10, 12, 8, 2, 5, 21, 8, 6, 2, 3
A) 7 B) 6.5 C) 2 D) 19

343) A entertainer pulls colored ribbons out of a box at random for a dance routine. The box contains 5 red ribbons and 6 blue ribbons. The other ribbons in the box are green. If a ribbon is pulled out of the box at random, the probability that the ribbon is red is $1/3$. How many green ribbons are in the box?
A) 3 B) 4 C) 5 D) 6

344) An athlete ran 10 miles in 1.5 hours. The graph below shows the miles the athlete ran every 10 minutes. According to the graph, how many miles did the athlete run in the first 30 minutes?

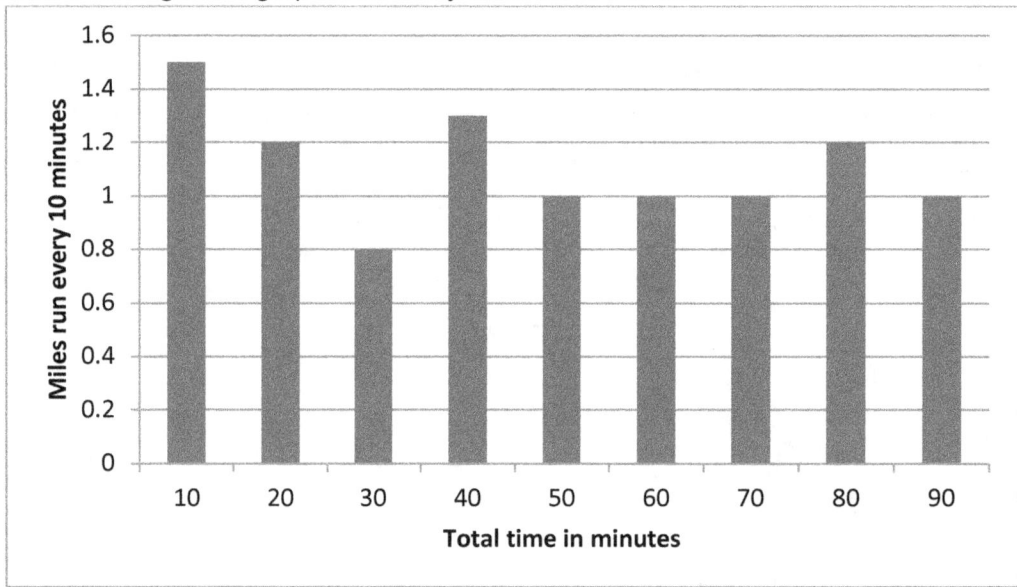

A) 0.8 miles
B) 2.0 miles
C) 3.0 miles
D) 3.5 miles

345) The residents of Hendersonville took a census. As part of the census, each resident had to indicate how many relatives they had living within a ten-mile radius of the town. The results of that particular question on the census are represented in the graph below.

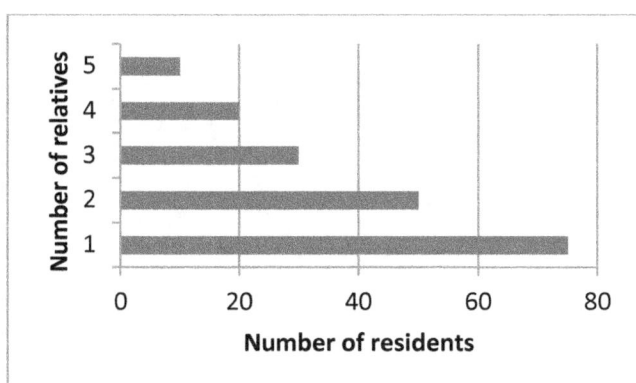

How many residents of Hendersonville had more than 3 relatives living within a ten-mile radius of the town?
A) 10　　　　　　　　B) 20　　　　　　　　C) 30　　　　　　　　D) 155

346) An illusionist has a box of pieces of colored rope for an illusion that he performs at a live show. The box contains 4 pieces of blue rope, 2 pieces of white rope, 1 piece of green rope, 4 pieces of yellow rope, and 5 pieces of black rope. The illusionist selects pieces of rope at random and the first piece of rope he selects is blue. What is the probability that he will select a piece of blue rope again on the second draw? Note that the pieces of rope are not put back into the box once they have been selected.

A) $1/5$　　　　　　　B) $1/4$　　　　　　　C) $3/16$　　　　　　D) $4/15$

347) Mr. Smith teaches a class of 25 students. Ten of the students in his class participate in drama club. In which graph below does the dark gray area represent the percentage of students who participate in drama club?

A)
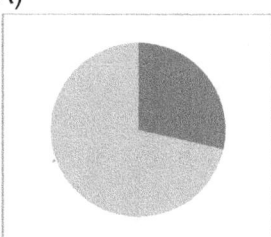

B)
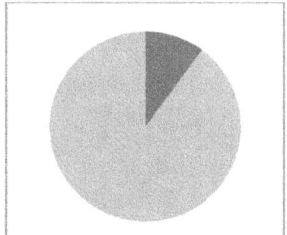

C)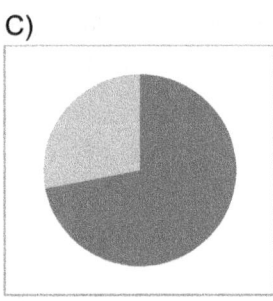

D)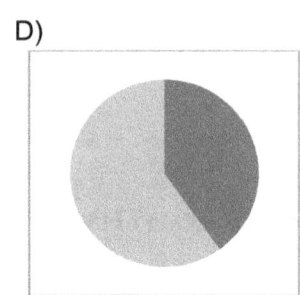

348) In Brown County Elementary School, parents are advised to have their children vaccinated against five childhood diseases. According to the chart below, how many children were vaccinated against at least three diseases?

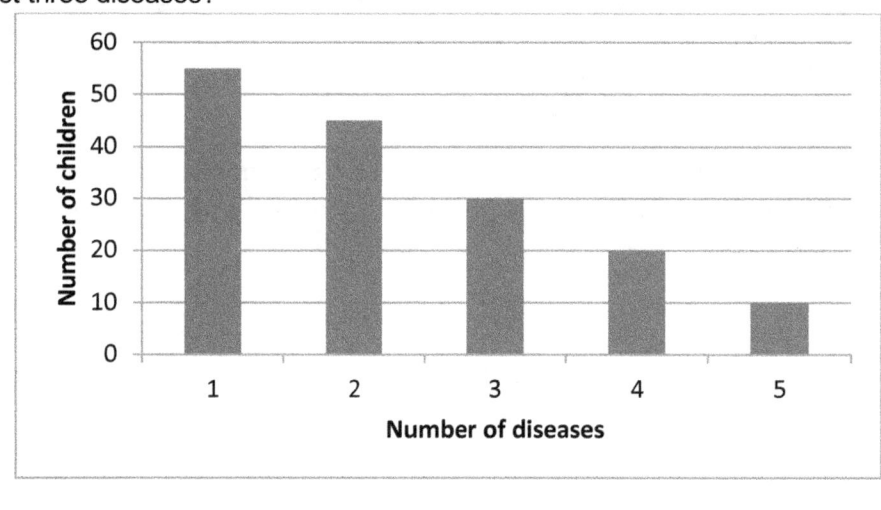

A) 30   B) 50   C) 60   D) 100

349) A doctor measures the pulse for several patients one morning. She recorded these results: 54, 68, 62, 60, 75, 58, 84, and 91. What is the range for this group of patients?

A) 30   B) 37   C) 65   D) 69

350) A wastewater company measures the amount of wastewater usage per household in wastewater units (WWU's). During one calendar quarter, the houses on a particular street had these measurements: 682, 534, 689, 783, and 985. What is the mode of wastewater usage in WWU's for this quarter for these properties?

A) no mode   B) 451   C) 689   D) 734.6

Algebra, Measurement, and Estimation

351) Perform the operations: $(5x - 2)(3x^2 + 5x - 8)$
A) $15x^3 + 19 + 50x + 16$
B) $15x^3 + 19x^2 + 70x - 16$
C) $15x^3 + 19x^2 - 50x + 16$
D) $15x^3 + 19x^2 - 70x + 16$

352) Solve for x and y: $x + 5y = 24$ and $8x + 2y = 40$
A) (4, 4)   B) (–4, 4)   C) (40, 4)   D) (4, 38)

353) Perform the operation and express as one fraction: $\dfrac{2}{10x} + \dfrac{3}{12x^2}$

A) $\dfrac{30x}{24x^2}$
B) $\dfrac{5}{10x+12x^2}$
C) $\dfrac{4x+5}{20x^2}$
D) $\dfrac{24x^2}{30x}$

354) $\sqrt{50} + 4\sqrt{32} + 7\sqrt{2} = ?$
A) $8\sqrt{58}$ B) $28\sqrt{2}$ C) $15\sqrt{58}$ D) $16\sqrt{2}$

355) $10a^2b^3c \div 2ab^2c^2 = ?$
A) $5c \div ab$ B) $5a \div bc$ C) $5ab \div c$ D) $5ac \div b$

356) $\dfrac{\sqrt{48}}{3} + \dfrac{5\sqrt{5}}{6} = ?$

A) $\dfrac{4\sqrt{3}+5\sqrt{5}}{6}$
B) $\dfrac{8\sqrt{3}+5\sqrt{5}}{6}$
C) $\dfrac{\sqrt{48}+5\sqrt{5}}{9}$
D) $\dfrac{6\sqrt{48}+5\sqrt{5}}{18}$

357) What is the value of $\dfrac{x-3}{2-x}$ when $x = 1$?

A) 2   B) –2   C) ½   D) –½

358) $\sqrt[3]{5} \times \sqrt[3]{7} = ?$
A) $\sqrt[3]{13}$ B) $\sqrt[6]{13}$ C) $\sqrt[9]{13}$ D) $\sqrt[3]{35}$

359) If x and y are positive integers, the expression $\dfrac{1}{\sqrt{x}-\sqrt{y}}$ is equivalent to which of the following?

A) $\sqrt{x}-y$ B) $\sqrt{x}+y$ C) $\dfrac{\sqrt{x}-y}{1}$ D) $\dfrac{\sqrt{x}+\sqrt{y}}{x-y}$

360) If $x + y = 5$ and $a + b = 4$, what is the value of $(3x + 3y)(5a + 5b)$?
A) 9   B) 35   C) 200   D) 300

361) What are two possible values of x for the following equation? $x^2 + 6x + 8 = 0$
A) 1 and 2
B) 2 and 4
C) 6 and 8
D) –2 and –4

362) Which of the following mathematical expressions equals $3/xy$?
A) $3/x \times 3/y$
B) $3 \div 3xy$
C) $3 \div (xy)$
D) $1/3 \div 3xy$

363) If $\frac{1}{5}x + 3 = 5$, then $x = ?$
A) $\frac{8}{5}$
B) $-\frac{8}{5}$
C) 8
D) 10

364) Solve for x: $x^2 - 12x + 35 < 0$
A) 5 > x > 7
B) 5 > x or x < 7
C) 5 < x < 7
D) 5 < x or x > 7

365) Which of the following expressions is equivalent to $2xy - 8x^2y + 6y^2x^2$?
A) $2(xy - 4x^2y + 3x^2y^2)$
B) $2xy(-4x + 3xy)$
C) $2xy(1 - 4x + 3xy)$
D) $2xy(1 + 4x - 3xy)$

366) Look at the scatterplot below and then choose the best answer from the options that follow.

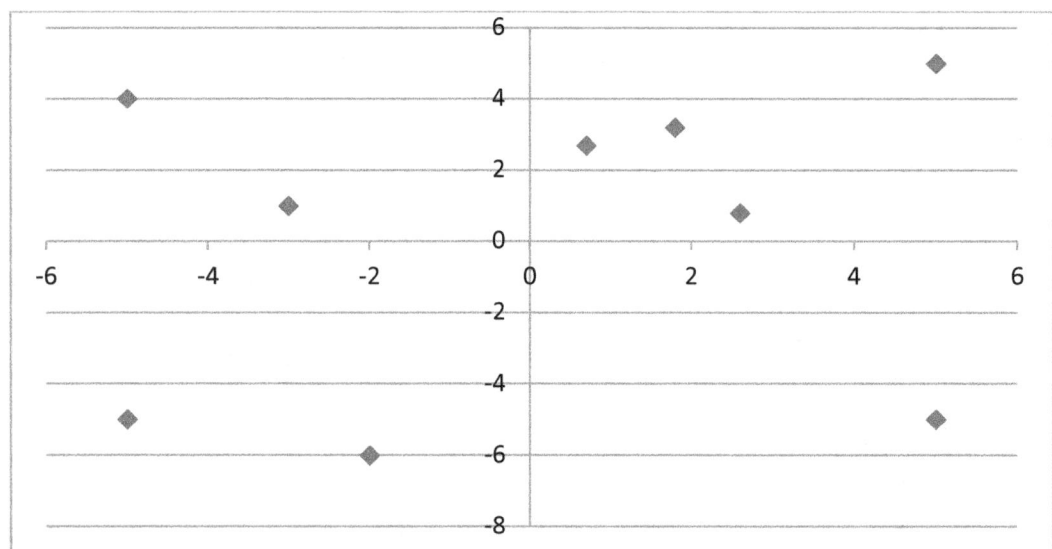

A) The scatterplot suggests a strong positive linear relationship between x and y.
B) The scatterplot suggests a strong negative linear relationship between x and y.
C) The scatterplot suggests a weak positive linear relationship between x and y.
D) The scatterplot suggests that there is no relationship between x and y.

367) The line on the xy-graph below forms the diameter of the circle. What is the approximate circumference of the circle?

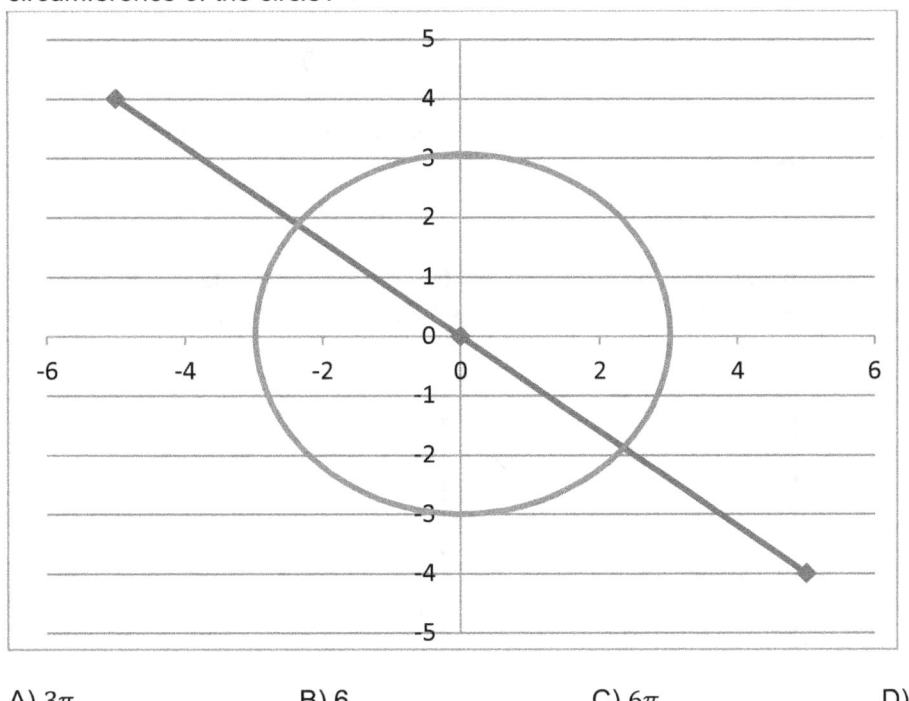

A) 3π B) 6 C) 6π D) 9

368) Mr. Lee is going to build a new garage. The garage will have a square base and a pyramid-shaped roof. The base measurement of the interior of the garage is 20 feet. The height of the interior of the garage is 18 feet. The height of the roof from the center of its base to its peak is 15 feet. A diagram of the garage is shown below:

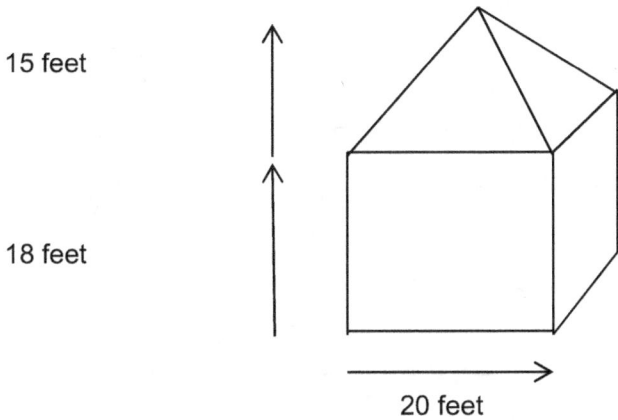

What fraction expresses the ratio of the volume of the roof of the garage to the volume of the interior of the garage?

A) $5/6$ B) $5/18$ C) $1/4$ D) $3/4$

369) A large wheel (L) has a radius of 10 inches. A small wheel (S) has a radius of 6 inches. If the large wheel is going to travel 360 revolutions, how many more revolutions does the small wheel need to make to cover the same distance?
A) 120
B) 240
C) 360
D) 720

370) The vertex of an angle is at the center of a circle. If the angle measures 20 degrees, and the diameter of the circle is 2, what is the arc length relating to the angle?
A) $\pi/9$
B) $\pi/18$
C) 9π
D) 18π

371) A farmer has a rectangular pen in which he keeps animals. He has decided to divide the pen into two parts. To divide the pen, he will erect a fence diagonally from the two corners, as shown in the diagram below. How long in yards is the diagonal fence?

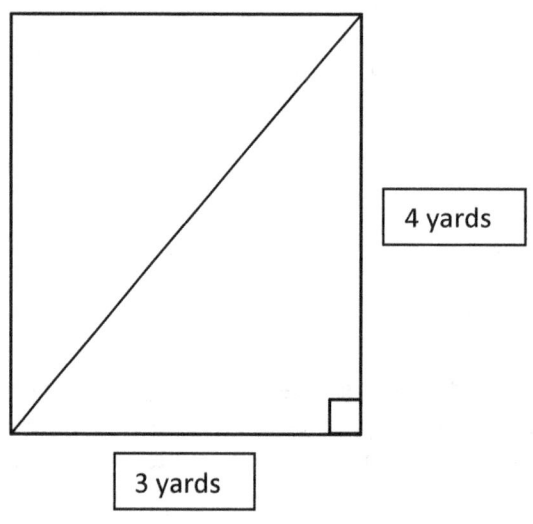

A) 4
B) 5
C) 5.5
D) 6

372) What is the area of the figure below?

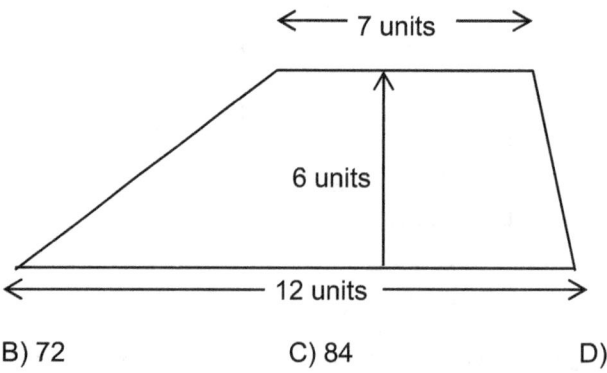

A) 57
B) 72
C) 84
D) 202

373) The illustration below shows a pentagon. The shaded part at the top of the pentagon has a height of 6 inches.

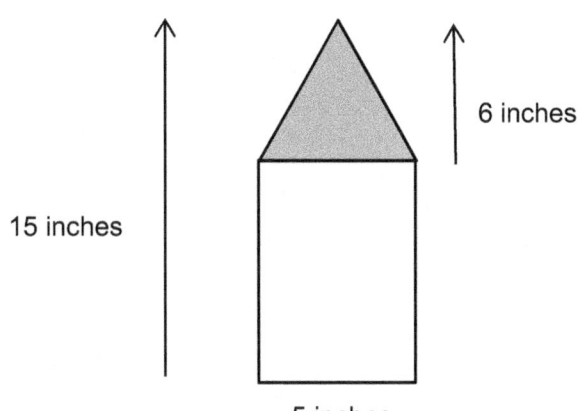

5 inches

The height of the entire pentagon is 15 inches, and the base of the pentagon is 5 inches. What is the area of the entire pentagon?

A) 15   B) 30   C) 45   D) 60

374) What equation can be used for a 90° angle?

A)   π × radian × 2   B)   π × radian × 4   C)   (π ÷ 2) × radian   D)   (π × 2) ÷ radian

375) The area of a square is 64 square units. This square is made up of smaller squares that measure 4 square units each. How many of the smaller squares are needed to make up the larger square?
A)   8   B) 12   C) 16   D) 24

376) Which of the following statements about parallelograms is true?
A)   A parallelogram has no right angles.
B)   A parallelogram has opposite angles which are congruent.
C)   The opposite sides of a parallelogram are unequal in measure.
D)   A rectangle is not a parallelogram.

377) Which of the following statements best describes supplementary angles?
A)   Supplementary angles must add up to 90 degrees.
B)   Supplementary angles must add up to 180 degrees.
C)   Supplementary angles must add up to 360 degrees.
D)   Supplementary angles must be congruent angles.

378) A is 3 times B, and B is 3 more than 6 times C. Which of the following describes the relationship between A and C?
A)   A is 9 more than 18 times C.   B) A is 3 more than 3 times C.
C)   A is 3 more than 18 times C.   C) A is 6 more than 3 times C.

379) The graph below illustrates which of the following functions?

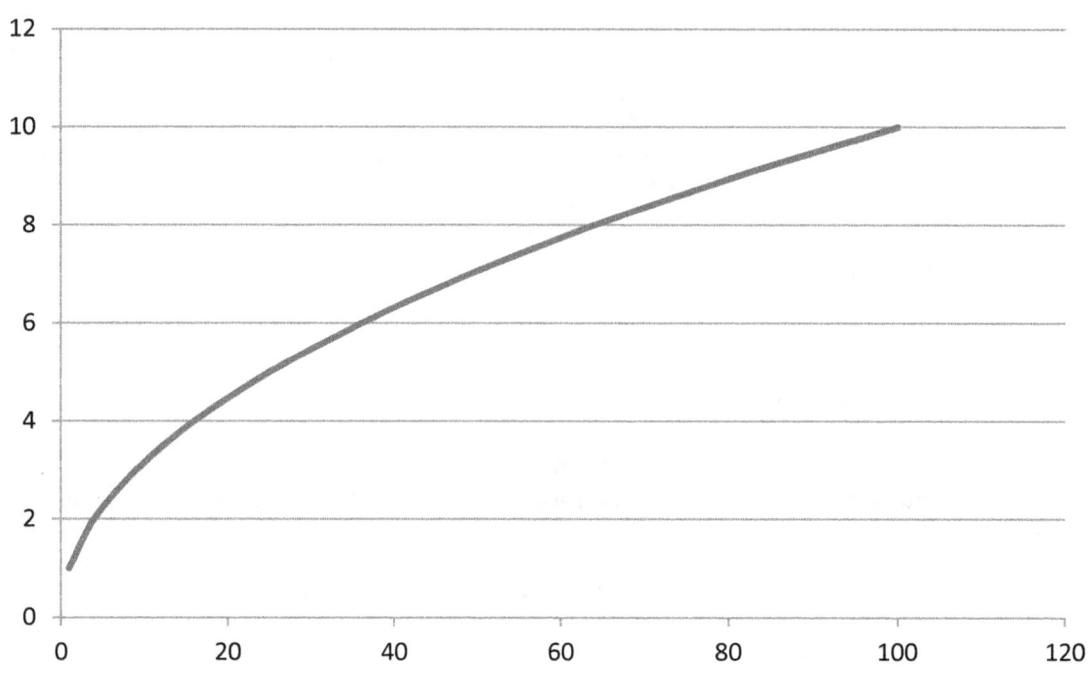

A) $f(x) = \frac{1}{x}$ B) $f(x) = 2^x$ C) $f(x) = \sqrt{x}$ D) $f(x) = x^2$

380) |6 – 13| = ?
A) 19   B) 7   C) –7   D) –19

381) Express 81 as a logarithmic function.
A) $81 = \log_2 9$
B) $2 = \log_9 81$
C) $9 = \log_2 81$
D) $81 = \log_9 2$

382) If $f_2(x) = \sqrt{x} + 3$ and $f_1(x) = 3x + 1$, what is the value of $f_1(f_2(9))$?
A) $\sqrt{28} + 3$
B) 19
C) 28
D) 6

383) The street that runs from Bilal's house (B) to Yoko's house (Y) is at a 60° angle. If Bilal's house is 2 miles from the grocery store (G), what equation can be used to calculate the distance from Yoko's house to the grocery store?

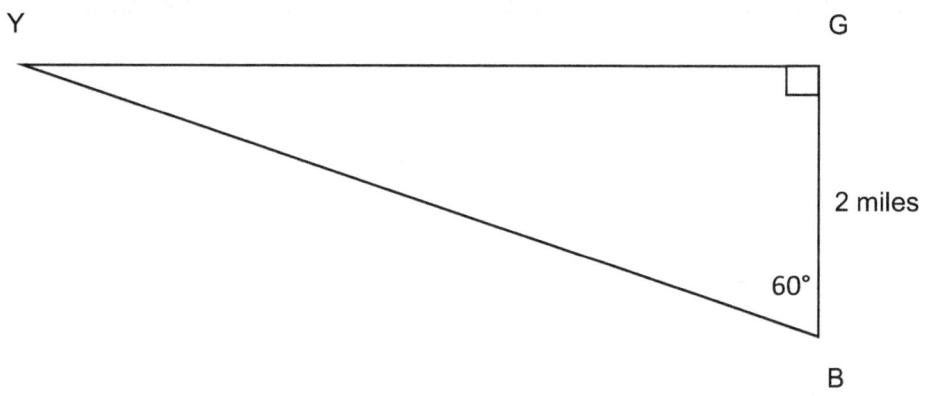

A) 2 × cos60°   B) 2 ÷ tan60°   C) 2 × tan60°   D) 2 × sin60°

384) The sine of a 72 degree angle is equal to which of the following?

A) 72° − 45°   B) the cosine of 18°   C) the tangent of 18°   D) (0.951056516)²

385) Angle A is a 45 degree angle. The cosine of A is 0.70710678 and the tangent of angle A is 1. What is the sine of angle A?

A) 0.70710678   B) 0.292893   C) $\dfrac{1}{0.70710678}$ D) (0.70710678)²

386) Express in degrees: π ÷ 2 × radian
A) 45°   B) 90°   C) 180°   D) 360°

387) Find the midpoint between the following coordinates: (5, 7) and (11, −3)
A) (2, 5)   B) (5, 2)   C) (2, 8)   D) (8, 2)

388) If $\sqrt{9z + 18} = 9$, then z = ?
A) −1   B) 6   C) 7   D) 63

389) If $z = \dfrac{x}{1-y}$, then y = ?
A) $\dfrac{z}{x}$ B) $\dfrac{x}{z} - 1$ C) $-\dfrac{x}{z} + 1$ D) $z - zx$

390) Perform the operation: $\sqrt{6} \cdot (\sqrt{40} + \sqrt{6})$
A) $\sqrt{240} + \sqrt{6}$ B) $\sqrt{46} + 6$ C) 46   D) $4\sqrt{15} + 6$

391) Which of the following is equivalent to $a^{½}b^{¼}c^{¾}$?
A) $a^2bc^3 \div 4$ B) $4(a^2bc^3)$ C) $\sqrt{a} \times \sqrt[4]{b} \times \sqrt[4]{c^3}$ D) $(ab^{¼}c^{¾} \div 2)$

392) $ab^8 \div ab^2 = ?$

A) ab^6 B) ab^4 C) a^2b^6 D) b^6

393) In the following figure, which of the following points could be the intersection of the circle and the line $y = 5$?

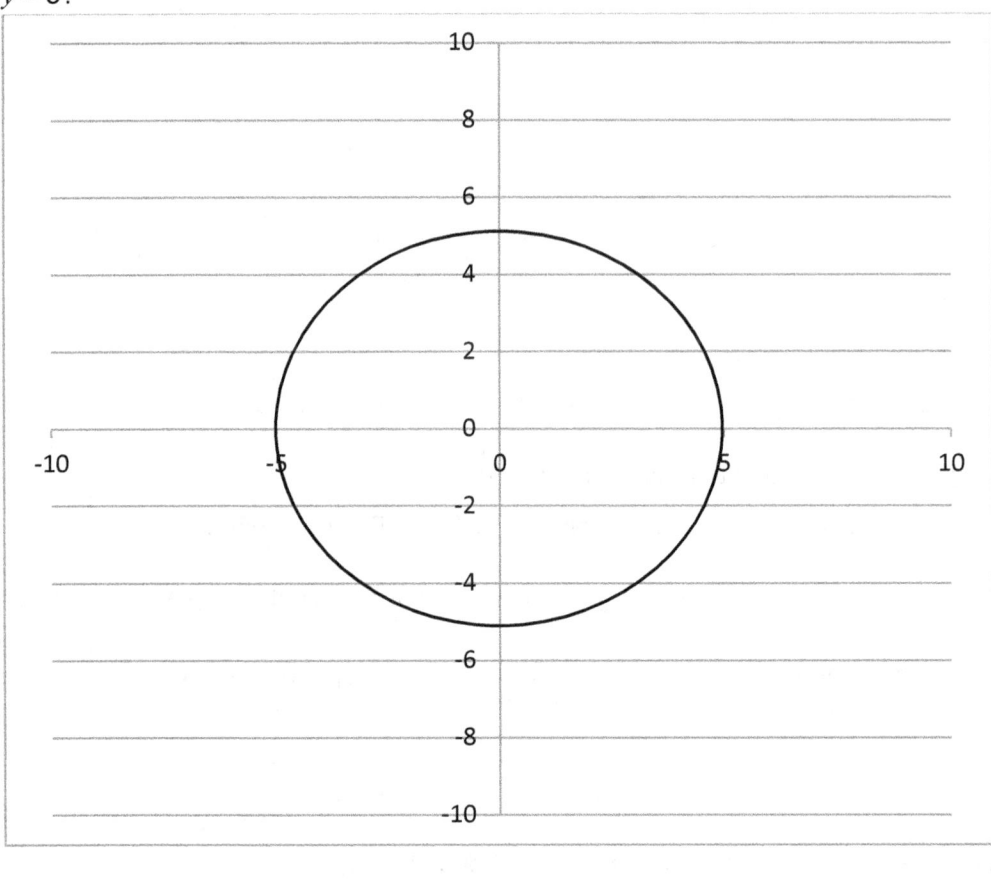

A) (0, 5) B) (5, 0) C) (0, –5) D) (–5, 0)

394) Martin is using a plaster mix to repair a wall in his house. The instructions for the mix state that 3 ounces of water should be added for every 4 ounces of plaster pounder used. If he uses 14 ounces of plaster powder, how much water should he add?
A) 56 ounces of water
B) 41 ounces of water
C) 10.5 ounces of water
D) 3 ounces of water

395) The graph below shows the cost in dollars of Item A as a function of the number of pounds purchased. The equation C = p × (7 ÷ 2) represents the cost of Item B, where p is the number of pounds and C is the cost. Which of the following statements best describes the relationship between the cost of Item A and the cost of Item B?

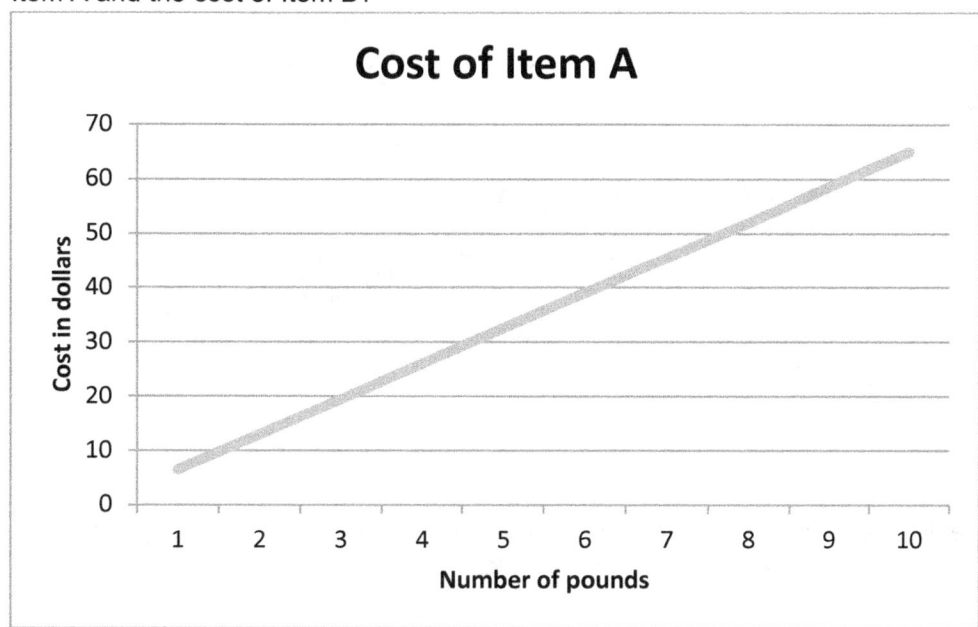

A) Item B costs $3.50 less per pound than Item A.
B) Item B costs $3.50 more per pound than Item A.
C) Item B costs $3 less per pound than Item A.
D) Item B costs $3 more per pound than Item A.

396) The graph of $y = f(x)$ is shown on the graph below.

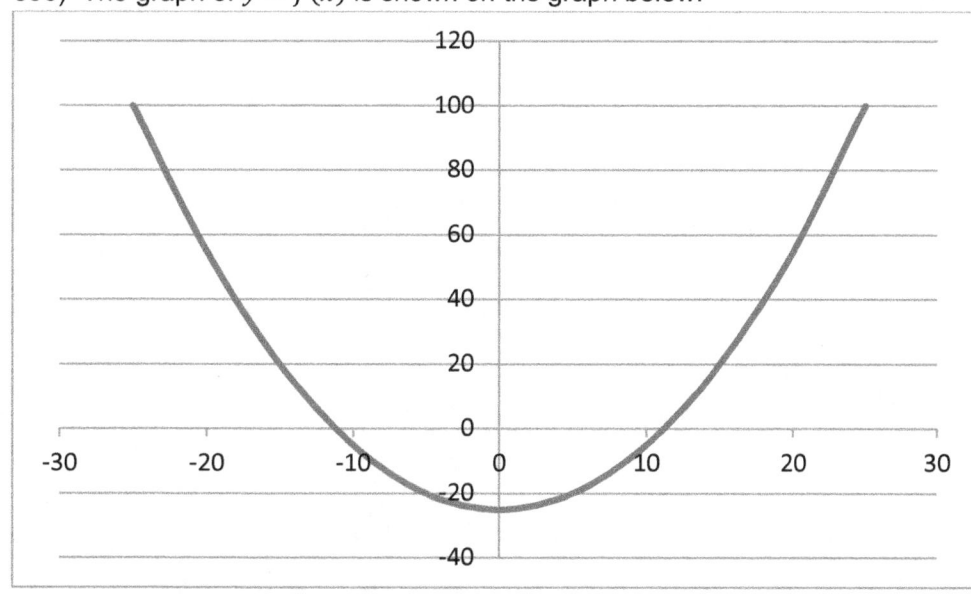

Which of the following could define $f(x)$?

A) $f(x) = \frac{x^2}{5} - 25$ B) $f(x) = 5x - 25$ C) $f(x) = 5x^2$ D) $f(x) = \frac{x}{25}$

397) In the function $f(x) = w(x + 4)(x - 2)^z$, w and z are integer constants and z is positive. The end behaviour of the graph of $y = f(x)$ is negative for large positive values of x and positive for large negative values of x. Which of the following statements is true with respect to w and z?
A) w is negative and z is odd.
B) w is positive and z is odd.
C) w is negative and z is even.
D) w is positive and z is even.

398) Perform the operation and express as one fraction: $\dfrac{1}{a+1} + \dfrac{1}{a}$

A) $\dfrac{2}{2a+1}$
B) $\dfrac{a+1}{a}$
C) $\dfrac{a^2+a}{2a+1}$
D) $\dfrac{2a+1}{a^2+a}$

399) For all $x \neq 0$ and $y \neq 0$, $\dfrac{5x}{1/xy} = ?$

A) $\dfrac{5x}{xy}$
B) $\dfrac{xy}{5x}$
C) $\dfrac{5}{y}$
D) $5x^2y$

400) The speed of a rocket is 25,000 miles per hour. How far does the rocket travel in 108,000 seconds?
A) 7.5×10^4 miles
B) 7.5×10^5 miles
C) 7.5×10^6 miles
D) 4.2×10^6 miles

Solutions and Explanations for Practice Test Set 1 – Questions 1 to 80

1) The correct answer is D. Because two negatives make a positive, we know that – (–6) = 6. So, we can substitute this into the equation in order to solve it: – (–6) + 2 = 6 + 2 = 8

2) The correct answer is D. For problems that ask you to find the largest possible product of two even integers, first you need to divide the sum by 2. The sum in this problem is 30, so divide by 2: 30 ÷ 2 = 15. Now take the result from this division and find the 2 nearest even integers that are 1 number higher and lower: 15 + 1 = 16; 15 − 1 = 14. Finally, multiply these two numbers together in order to get the product: 16 × 14 = 224.

3) The correct answer is C. Multiply the numerators: 1 × 2 = 2. Then multiply the denominators: 5 × 3 = 15. These numbers form the new fraction: 2/15

4) The correct answer is B. Remember to invert the second fraction by putting the denominator on the top and the numerator on the bottom. So the second fraction $\frac{2}{3}$ becomes $\frac{3}{2}$ when inverted. Use the inverted fraction to solve the problem: $\frac{4}{7} \div \frac{2}{3} = \frac{4}{7} \times \frac{3}{2} = \frac{4 \times 3}{7 \times 2} = \frac{12}{14}$

5) The correct answer is A. STEP 1: To find the LCD, you have to look at the factors for each denominator. Factors are the numbers that equal a product when they are multiplied by each other. The factors of 8 are: 1 × **8** = 8; **2** × 4 = 8. The factors of 16 are: 1 × 16 = 16; **2** × **8** = 16; 4 × 4 = 16. STEP 2: Determine which factors are common to both denominators by comparing the lists of factors. In this problem, the factors of 2 and 8 are common to the denominators of both fractions. (The numbers in bold above are the common factors.) STEP 3: Multiply the common factors to get the lowest common denominator. The numbers that are in bold above are then used to calculate the lowest common denominator: 2 × 8 = 16. STEP 4: Convert the denominator of each fraction to the LCD. You convert the fraction by referring to the factors from step 3. Multiply the numerator and the denominator by the same factor. We convert the first fraction as follows: $\frac{1}{8} \times \frac{2}{2} = \frac{2}{16}$. We do not need to convert the second fraction because it already has the LCD. STEP 5: When both fractions have the same denominator, you can perform the operation to solve the problem: $\frac{2}{16} + \frac{3}{16} = \frac{5}{16}$

6) The correct answer is D. STEP 1: Look at the factors of the numerator and denominator. The factors of 12 are: 1 × 12 = 12; **2** × 6 = 12; 3 × 4 = 12. The factors of 14 are: 1 × 14 = 14; **2** × 7 = 14. So, we can see that the numerator and denominator have the common factor of 2. STEP 2: Simplify the fraction by dividing the numerator and denominator by the common factor. Simplify the numerator: 12 ÷ 2 = 6. Then simplify the denominator: 14 ÷ 2 = 7. STEP 3: Use the results from step 2 to form the new fraction. The numerator from step 2 is 6. The denominator is 7. So, the new fraction is $\frac{6}{7}$.

7) The correct answer is C. STEP 1: Convert the first mixed number to an integer plus a fraction. $4\frac{1}{8} = 4 + \frac{1}{8}$. STEP 2: Then multiply the integer by a fraction whose numerator and denominator are the same as the denominator of the existing fraction: $4 + \frac{1}{8} = \left(4 \times \frac{8}{8}\right) + \frac{1}{8} = \frac{4 \times 8}{8} + \frac{1}{8} = \frac{32}{8} + \frac{1}{8}$

STEP 3: Add the two fractions to get your new fraction: $\frac{32}{8} + \frac{1}{8} = \frac{33}{8}$. Convert the second mixed number to a fraction, using the same steps that we have just completed for the first mixed number: $3\frac{5}{6} = 3 + \frac{5}{6} = \left(3 \times \frac{6}{6}\right) + \frac{5}{6} = \frac{18}{6} + \frac{5}{6} = \frac{23}{6}$

Now that you have converted both mixed numbers to fractions, find the lowest common denominator and subtract to solve.

$\frac{33}{8} - \frac{23}{6} = \left(\frac{33}{8} \times \frac{3}{3}\right) - \left(\frac{23}{6} \times \frac{4}{4}\right) = \frac{(33 \times 3)}{(8 \times 3)} - \frac{(23 \times 4)}{(6 \times 4)} = \frac{99}{24} - \frac{92}{24} = \frac{99 - 92}{24} = \frac{7}{24}$

8) The correct answer is A. Do the multiplication and division from left to right. So, take the number to the left of the multiplication or division symbol and multiply or divide that number by the number on the right of the symbol. We need to multiply –5 by 4 and then divide 6 by 3. You can see the order of operations more clearly if you put in parentheses to group the numbers together.
–5 × 4 – 6 ÷ 3 = (–5 × 4) – (6 ÷ 3) = –20 – 2 = –22

9) The correct answer is B. For this type of problem, do the operations inside the **parentheses** first.
$\frac{4 \times (5-2)^3 + 6}{7 - 4 \div 2} = \frac{4 \times (3)^3 + 6}{7 - 4 \div 2}$

Then do the operation on the **exponent**.
$\frac{4 \times (3)^3 + 6}{7 - 4 \div 2} = \frac{4 \times 27 + 6}{7 - 4 \div 2}$

Then do the **multiplication** and **division**.
$\frac{4 \times 27 + 6}{7 - 4 \div 2} = \frac{(4 \times 27) + 6}{7 - (4 \div 2)} = \frac{108 + 6}{7 - 2}$

Then do the **addition** and **subtraction**.
$\frac{108 + 6}{7 - 2} = \frac{114}{5}$

10) The correct answer is C. Determine the dollar amount of the reduction or discount:
$60 original price – $45 sale price = $15 discount. Then divide the discount by the original price to get the percentage of the discount: $15 ÷ $60 = 0.25 = 25%

11) The correct answer is B. The sales price of each item is five times the cost. The cost is expressed as B, so the sales price is 5B. The difference between the sales price of item and the cost of each item is the profit:
Sales Price − Cost = Profit
5B − B = 4B

12) The correct answer is D. The thousandths place is the third one to the right of the decimal. So, 0.96547 rounded to the nearest thousandth is 0.965.

13) The correct answer is B. Perform the division, and then check the decimal places of the numbers. Divide as follows: 1523.48 ÷ 100 = 15.2348. Reading our result from left to right: 1 is in the tens place, 5 is in the ones place, 2 is in the tenths place, 3 is in the hundredths place, 4 is in the thousandths place, and 8 is in the ten-thousandths place.

14) The correct answer is C. Determine which numbers can be calculated by multiplying another integer by 3: 1 × 3 = 3; 2 × 3 = 6; 3 × 3 = 9; and 4 × 3 = 12. 12 is greater than 11, so the nearest product to 11 from the list above is 9. Finally, we subtract these two numbers to get the remainder: 11 − 9 = 2

15) The correct answer is A. Line up the decimal points as shown when adding. Always remember to "carry the 1."

```
  1 1
  4.25
  0.003
  0.148
  4.401
```

16) The correct answer is A. From the facts in the problem, we know that η needs to be greater than $^{25}/_{13}$. If we convert $^{25}/_{13}$ to decimal form, we get 1.923077. The square root of 36 is 6, so (A) is the correct response because it is greater than 1.923077.

17) The correct answer is B. $7x$ is between 5 and 6, so set up an inequality as follows: $5 < 7x < 6$
Then insert the fractions from the answer choices for the value of x to solve the problem.
$5 < (7 \times {}^3/_4) < 6$
$5 < [(7 \times 3) \div 4] < 6$
$5 < (21 \div 4) < 6$
$5 < 5.25 < 6$
5.25 is between 5 and 6, so $^3/_4$ is the correct answer.

18) The correct answer is D. First of all, you need to determine the difference in temperature during the entire time period: 62 − 38 = 24 degrees less. Then calculate how much time has passed. From 5:00 PM to 11:00 PM, 6 hours have passed. Next, divide the temperature difference by the amount of time that has passed to get the temperature change per hour: 24 degrees ÷ 6 hours = 4 degrees less per hour. To calculate the temperature at the stated time, you need to calculate the time difference. From 5:00 PM to 9:00 PM, 4 hours have passed. So, the temperature difference during the stated time is 4 hours × 4 degrees per hour = 16 degrees less. Finally, deduct this from the beginning temperature to get your final answer: 62° F − 16° F = 46° F.

19) The correct answer is C. For your first step, determine how many square feet there are in total: 2000 square feet per room × 8 rooms = 16,000 square feet in total. Then you need to divide by the coverage rate: 16,000 square feet to cover ÷ 900 square feet coverage per bucket = 17.77 buckets needed. It is not possible to purchase a partial bucket of paint, so 17.77 is rounded up to 18 buckets of paint.

20) The correct answer is D. Divide the distance traveled by the time in order to get the speed in miles per hour. Remember that in order to divide by a fraction, you need to invert the fraction, and then multiply. 3.6 miles ÷ $^3/_4$ = 3.6 × $^4/_3$ = (3.6 × 4) ÷ 3 = 14.4 ÷ 3 = 4.8 miles per hour

21) The correct answer is B. For your first step, add the subsets of the ratio together: 6 + 7 = 13. Then divide this into the total: 117 ÷ 13 = 9. Finally, multiply the result from the previous step by the subset of males from the ratio: 6 × 9 = 54 males in the class.

22) The correct answer is A. The outcome of an earlier roll does not affect the outcome of the next roll. When rolling a pair of dice, the possibility of an odd number is always $1/2$, just as the possibility of an even number is always $1/2$. We can prove this mathematically by looking at the possible outcomes:
1,1 1,2 1,3 1,4 1,5 1,6
2,1 2,2 2,3 2,4 2,5 2,6
3,1 3,2 3,3 3,4 3,5 3,6
4,1 4,2 4,3 4,4 4,5 4,6
5,1 5,2 5,3 5,4 5,5 5,6
6,1 6,2 6,3 6,4 6,5 6,6
The odd number combinations are highlighted:
1,1 **1,2** 1,3 **1,4** 1,5 **1,6**
2,1 2,2 **2,3** 2,4 **2,5** 2,6
3,1 **3,2** 3,3 **3,4** 3,5 **3,6**
4,1 4,2 **4,3** 4,4 **4,5** 4,6
5,1 **5,2** 5,3 **5,4** 5,5 **5,6**
6,1 6,2 **6,3** 6,4 **6,5** 6,6
So, we can see that an odd number will be rolled half of the time.

23) The correct answer is D. To find the mean, add up all of the items in the set and then divide by the number of items in the set. Here we have 7 numbers in the set, so we get our answer as follows: (89 + 65 + 75 + 68 + 82 + 74 + 86) ÷ 7 = 539 ÷ 7 = 77

24) The correct answer is A. The mode is the number that occurs the most frequently in the set. Our data set is: 1, 1, 3, 2, 4, 3, 1, 2, 1. The number 1 occurs 4 times in the set, which is more frequently than any other number in the set, so the mode is 1.

25) The correct answer is B. The problem provides the number set: 8.19, 7.59, 8.25, 7.35, 9.10
First of all, put the numbers in ascending order: 7.35, 7.59, 8.19, 8.25, 9.10. Then find the one that is in the middle: 7.35, 7.59, **8.19**, 8.25, 9.10

26) The correct answer is C. To calculate the range, the low number in the set is deducted from the high number in the set. The problem set is: 98.5, 85.5, 80.0, 97, 93, 92.5, 93, 87, 88, 82. The high number is 98.5 and the low number is 80, so the range is 18.5 (98.5 – 80 = 18.5).

27) The correct answer is B.
Step 1 – Calculate the arithmetic mean for the data set.
99 + 98 + 74 + 69 + 87 + 83 = 510 total for all six students
510 divided by 6 students equals an arithmetic mean of 85 for the group.
Step 2 – Find the "difference from the mean" for each item in the data set by subtracting the mean from each value.
Student 1: 99 – 85 = 14
Student 2: 98 – 85 = 13
Student 3: 74 – 85 = –11
Student 4: 69 – 85 = –16
Student 5: 87 – 85 = 2
Student 6: 83 – 85 = –2

Step 3 – Square the "difference from the mean" for each item in the data set.

Student 1: $14^2 = 196$
Student 2: $13^2 = 169$
Student 3: $-11^2 = 121$
Student 4: $-16^2 = 256$
Student 5: $2^2 = 4$
Student 6: $-2^2 = 4$

Step 4 – Calculate the mean of the squared figures to calculate the variance.

Variance = 196 + 169 + 121 + 256 + 4 + 4 = 750 ÷ 6 = 125

28) The correct answer is C. The standard deviation of a data set measures the spread of the data around the mean of the data set. The standard deviation is calculated by taking the square root of the variance. So, we use the variance calculated in the previous question: 125.
The square root is $\sqrt{125} = \sqrt{5 \times 5 \times 5} = 5\sqrt{5}$

29) The correct answer is C. The question is asking for the union of A with the intersection of B and C. Since (B ∩ C) is in parentheses, we need to find that intersection first. Look at sets B and C and highlight the numbers that they have in common: B = {4, 8, **12**, **16**, **20**}; C = {**12**, 14, **16**, 18, **20**}, so (B ∩ C) = {12, 16, 20}. Set A before the union with this intersection was A = {5, 10, 15, 20, 25}. The number 20 is already in set A, so add the other 2 numbers from (B ∩ C) to set A to get your answer:
{5, 10, 12, 15, 16, 20, 25}.

30) The correct answer is C. At the beginning of the year, 15% of the 1,500 creatures were fish, so there were 225 fish at the beginning of the year (1,500 × 0.15 = 225). In order to find the percentage of fish at the end of the year, we first need to add up the percentages for the other animals: 40% + 23% + 21% = 84%. Then subtract this amount from 100% to get the remaining percentage for the fish: 100% – 84% = 16%. Multiply the percentage by the total to get the number of fish at the end of the year: 1,500 × 0.16 = 240. Then subtract the beginning of the year from the end of the year to calculate the increase in the number of fish: 240 – 225 = 15.

31) The correct answer is D. The factors of 50 are: 1 × 50 = 50; 2 × 25 = 50; 5 × 10 = 50. If any of your factors are perfect squares, you can simplify the radical. 25 is a perfect square, so, you need to factor inside the radical sign as shown to solve the problem: $\sqrt{50} = \sqrt{25 \times 2} = \sqrt{5^2 \times 2} = \sqrt{5^2} \times \sqrt{2} = 5\sqrt{2}$

32) The correct answer is D. 36 is the common factor, So, factor the amounts inside the radicals and simplify: $\sqrt{36} + 4\sqrt{72} - 2\sqrt{144} = \sqrt{36} + 4\sqrt{36 \times 2} - 2\sqrt{36 \times 4} =$
$\sqrt{6 \times 6} + 4\sqrt{(6 \times 6) \times 2} - 2\sqrt{(6 \times 6) \times 4} = 6 + (4 \times 6)\sqrt{2} - (2 \times 6)\sqrt{4} =$
$6 + 24\sqrt{2} - (12 \times 2) = 6 + 24\sqrt{2} - 24 = -18 + 24\sqrt{2}$

33) The correct answer is A. $\sqrt{7} \times \sqrt{11} = \sqrt{7 \times 11} = \sqrt{77}$

34) The correct answer is B. The cube root is the number which satisfies the equation when multiplied by itself two times: $\sqrt[3]{\frac{216}{27}} = \sqrt[3]{\frac{6 \times 6 \times 6}{3 \times 3 \times 3}} = \frac{6}{3} = 2$

35) The correct answer is A. The base number is 7. Add the exponents: $7^5 \times 7^3 = 7^{(5+3)} = 7^8$

36) The correct answer is B. The base is xy. Subtract the exponents: $xy^6 \div xy^3 = xy^{(6-3)} = xy^3$

37) The correct answer is B. We have the base number of 10 and we are multiplying, so we can add the exponent of 5 to the exponent of −1: (1.7 × 10⁵ miles per hour) × (2 × 10⁻¹ hours) = 1.7 × 2 × 10$^{(5 + -1)}$ = 3.4 × 10⁴ = 3.4 × 10,000 = 34,000 miles

38) The correct answer is D. When you have a fraction as an exponent, the numerator is new exponent and the denominator goes in front as the root: $\sqrt[7]{x^5} = \left(\sqrt[7]{x}\right)^5$

39) The correct answer is B. $x^{-5} = \frac{1}{x^5}$

40) The correct answer is C. We have a non-zero number raised to the power of zero, so it is equal to 1.

41) The correct answer is C.
$$\frac{b + \frac{2}{7}}{\frac{1}{b}} = \left(b + \frac{2}{7}\right) \div \frac{1}{b} = \left(b + \frac{2}{7}\right) \times \frac{b}{1} = b\left(b + \frac{2}{7}\right) = b^2 + \frac{2b}{7}$$

42) The correct answer is D. Find the lowest common denominator for the second fraction. Then add the numerators.
$$\frac{x^2}{x^2 + 2x} + \frac{8}{x} = \frac{x^2}{x^2 + 2x} + \left(\frac{8}{x} \times \frac{x + 2}{x + 2}\right) = \frac{x^2}{x^2 + 2x} + \frac{8x + 16}{x^2 + 2x} = \frac{x^2 + 8x + 16}{x^2 + 2x}$$

43) The correct answer is A. Multiply as shown: $\frac{2a^3}{7} \times \frac{3}{a^2} = \frac{2a^3 \times 3}{7 \times a^2} = \frac{6a^3}{7a^2}$

Then find the greatest common factor and cancel out to simplify: $\frac{6a^3}{7a^2} = \frac{6a \times a^2}{7 \times a^2} = \frac{6a \times \cancel{a^2}}{7 \times \cancel{a^2}} = \frac{6a}{7}$

44) The correct answer is B. Invert and multiply.
$$\frac{8x + 8}{x^4} \div \frac{5x + 5}{x^2} = \frac{8x + 8}{x^4} \times \frac{x^2}{5x + 5} = \frac{(8x \times x^2) + (8 \times x^2)}{(x^4 \times 5x) + (x^4 \times 5)} = \frac{8x^3 + 8x^2}{5x^5 + 5x^4}$$

Then factor out (x + 1) from the numerator and denominator and cancel out:

$$\frac{8x^3 + 8x^2}{5x^5 + 5x^4} = \frac{(8x^2 \times x) + (8x^2 \times 1)}{(5x^4 \times x) + (5x^4 \times 1)} = \frac{8x^2(x+1)}{5x^4(x+1)} = \frac{8x^2 \cancel{(x+1)}}{5x^4 \cancel{(x+1)}} = \frac{8x^2}{5x^4}$$

Finally, factor out x^2 and cancel it out: $\frac{8x^2}{5x^4} = \frac{x^2 \times 8}{x^2 \times 5x^2} = \frac{\cancel{x^2} \times 8}{\cancel{x^2} \times 5x^2} = \frac{8}{5x^2}$

45) The correct answer is D. Use the FOIL method to expand the polynomial.
FIRST – Multiply the first term from the first set of parentheses by the first term from the second set of parentheses: (**x** + 4y)(**x** + 4y) = x × x = x²
OUTSIDE – Multiply the first term from the first set of parentheses by the second term from the second set of parentheses: (**x** + 4y)(x + **4y**) = x × 4y = 4xy
INSIDE – Multiply the second term from the first set of parentheses by the first term from the second set of parentheses: (x + **4y**)(**x** + 4y) = 4y × x = 4xy
LAST– Multiply the second term from the first set of parentheses by the second term from the second set of parentheses: (x + **4y**)(x + **4y**) = 4y × 4y = 16y²
Finally, we add all of the products together: x² + 4xy + 4xy + 16y² = x² + 8xy + 16y²

46) The correct answer is C. As the quantity of sugar increases, the amount of sleep also increases. A positive linear relationship therefore exists between the two variables. This is represented in chart C since the amount of sleep is greater when the amount of sugar consumed is higher.

47) The correct answer is C. We can see that the line does not begin on exactly on (5, 5), nor does it begin on (5, 9) because the first point is slightly below the horizontal line for y = 5. Therefore, we can rule out answers A and D. If we look at x = 20 on the graph, we can see that y = 18 at this point. We can express this as the function: $f(x) = x \times 0.9$. Putting in the values of x from chart (C), we get the following: 5 × 0.9 = 4.5; 10 × 0.9 = 9; 15 × 0.9 =13.5; 20 × 0.9 = 18. This is represented in table C.

48) The correct answer is D. If a term or variable is subtracted within the parentheses, you have to keep the negative sign with it when you multiply.
FIRST: (**x** – y)(**x** + y) = x × x = x²
OUTSIDE: (**x** – y)(x + **y**) = x × y = xy
INSIDE: (x – **y**)(**x** + y) = –y × x = –xy
LAST: (x – **y**)(x + **y**) = –y × y = –y²
SOLUTION: x² + xy + – xy – y² = x² – y²

49) The correct answer is A. First, Isolate the whole numbers.

$50 - \frac{3x}{5} \geq 41$

$(50 - 50) - \frac{3x}{5} \geq 41 - 50$

$-\frac{3x}{5} \geq -9$

Then get rid of the denominator on the fraction.

$-\frac{3x}{5} \geq -9$

$\left(5 \times -\frac{3x}{5}\right) \geq -9 \times 5$

$-3x \geq -9 \times 5$
$-3x \geq -45$
Then isolate the remaining whole numbers.
$-3x \geq -45$
$-3x \div 3 \geq -45 \div 3$
$-x \geq -45 \div 3$
$-x \geq -15$
Then deal with the negative number.
$-x \geq -15$
$-x + 15 \geq -15 + 15$
$-x + 15 \geq 0$
Finally, isolate the unknown variable as a positive number.
$-x + 15 \geq 0$
$-x + x + 15 \geq 0 + x$
$15 \geq x$
$x \leq 15$

50) The correct answer is D. Substitute values as shown: $x - 2 > 5$ and $y = x - 2$, so $y > 5$. If two wizfits are being purchased, we need to solve for $2y$:
$y \times 2 > 5 \times 2$
$2y > 10$

51) The correct answer is B. For quadratic inequality problems like this one, you need to factor the inequality first. The factors of -9 are: -1×9; -3×3; 1×-9. Because we do not have a term with only the x variable, we need factors that add up to zero, so factor as shown:
$x^2 - 9 < 0$
$(x + 3)(x - 3) < 0$
Then find values for x by solving each parenthetical for 0.
$(x + 3) = 0$
$(-3 + 3) = 0$
$x = -3$

$(x - 3) = 0$
$(3 - 3) = 0$
$x = 3$

So, $x > -3$ or $x < 3$
You can then check your work to be sure that you have the inequality signs pointing the right way.

Use -2 to check $x > -3$. Since $-2 > -3$ is correct, our proof should also be correct:
$x^2 - 9 < 0$
$-2^2 - 9 < 0$
$4 - 9 < 0$
$-5 < 0$ CORRECT

Use 4 to check for $x < 3$. Since $4 < 3$ is incorrect, our proof should also be incorrect.
$x^2 - 9 < 0$

$4^2 - 9 < 0$
$16 - 9 < 0$
$7 < 0$ INCORRECT

Therefore, we have checked that x > –3 or x < 3.

52) The correct answer is D. We know that the products of 12 are: 1 × 12 = 12; 2 × 6 = 12; 3 × 4 = 12. So, add each of the two factors together to solve the first equation: 1 + 12 = 13; 2 + 6 = 8; 3 + 4 = 7. (3, 4) solves both equations, so it is the correct answer.

53) The correct answer is C. The first term of the second equation is *x*. To eliminate the *x* variable, we need to multiply the second equation by 3 because the first equation contains 3x.
$x + 2y = 8$
$(3 \times x) + (3 \times 2y) = (3 \times 8)$
$3x + 6y = 24$
Now subtract the new second equation from the original first equation.
$3x + 3y = 15$
$\underline{-(3x + 6y = 24)}$
$-3y = -9$
Then solve for *y*.
$-3y = -9$
$-3y \div -3 = -9 \div -3$
$y = 3$
Using our original second equation of $x + 2y = 8$, substitute the value of 3 for *y* to solve for *x*.
$x + 2y = 8$
$x + (2 \times 3) = 8$
$x + 6 = 8$
$x + 6 - 6 = 8 - 6$
$x = 2$

54) The correct answer is D. Repeat the operation for each number from 3 to 5.
For x = 3:   x – 1 = 3 – 1 = 2
For x = 4:   x – 1 = 4 – 1 = 3
For x = 5:   x – 1 = 5 – 1 = 4
Then add the individual results together to get the answer: 2 + 3 + 4 = 9

55) The correct answer is A. When we compare the equations, we see that operation Ð is division: the number or variable immediately before Ð is is multiplied by 30; and the number or variable immediately after Ð is multiplied by 9. So, the new equation for (3 Ð y) becomes (30 × 3) ÷ (9 × y) = 10
$(30 \times 3) \div (9 \times y) = 10$
$90 \div 9y = 10$
$y = 1$

56) The correct answer is B. $x = \log_y Z$ is the same as $y^x = Z$. So, $2 = \log_8 64$ is the same as $8^2 = 64$. Check your answer by performing the operation on the number with the exponent: $8^2 = 8 \times 8 = 64$

57) The correct answer is C. Deduct the degrees provided for angle A from 180° to find out the total degrees of the two other angles: 180° − 32° = 148°. Since this is an isosceles triangle, the remaining two angles are have the same measurement. So, divide by two in order to find out how many degrees each angle has: 148° ÷ 2 = 74°

58) The correct answer is B. The angle given in the problem is 90°. If we divide the total of 360° in the circle by the 90° angle, we have: 360 ÷ 90 = 4. You can think of arc length as the partial circumference of a circle, so we can visualize that there are 4 such arcs along this circle. We can then multiply the number of arcs by the length of each arc to get the circumference of the circle: 4 × 8π = 32π (circumference). Finally, use the formula for the circumference of the circle to solve.
Circumference = π × radius × 2
32π = π × 2 × radius
32π ÷ 2 = π × 2 × radius ÷ 2
16π = π × radius
16 = radius

59) The correct answer is D. The area of a rectangle is equal to its length times its width. The field is 32 yards wide and 100 yards long, so now we can substitute the values.
rectangle area = width × length
rectangle area = 32 × 100
rectangle area = 3200

60) The correct answer is A. Substitute the value of the diameter into the formula:
circumference = D × π = 6π

61) The correct answer is D. The center of this circle is (−5, 5) and the point of tangency is (−5, 0). We need to subtract these two coordinates in order to find the length of the radius: (−5, 5) − (−5, 0) = (0, 5) In other words, the radius length is 5, so the diameter length is 10.

62) The correct answer is C. The base length of the triangle described in the problem, which is line segment YZ, is not given. So, we need to calculate the base length using the Pythagorean theorem. According to the Pythagorean theorem, the length of the hypotenuse is equal to the square root of the sum of the squares of the two other sides.
$\sqrt{A^2 + B^2} = C$
$\sqrt{4^2 + B^2} = 5$
$\sqrt{16 + B^2} = 5$
Now square each side of the equation in order to solve for the base length.
$\sqrt{16 + B^2} = 5$
$\left(\sqrt{16 + B^2}\right)^2 = 5^2$
$16 + B^2 = 25$
$16 - 16 + B^2 = 25 - 16$
$B^2 = 9$
$\sqrt{B^2} = \sqrt{9}$
$B = 3$

Now solve for the area of the triangle.
triangle area = (base × height) ÷ 2
triangle area = (3 × 4) ÷ 2
triangle area = 12 ÷ 2
triangle area = 6

63) The correct answer is D. Triangle XYZ is a 30° / 60°/ 90° triangle. Using the Pythagorean theorem, its sides are therefore in the ratio of 1: √3: 2. Using relative measurements, the line segment opposite the 30° angle is 1 unit long, the line segment opposite the 60° angle is √3 units long, and the line segment opposite the right angle (the hypotenuse) is 2 units long. In this problem, line segment XY is opposite the 30° angle, so it is 1 proportional unit long. Line segment YZ is opposite the 60° angle, so it is √3 proportional units long. Line segment XZ (the hypotenuse) is the line opposite the right angle, so it is 2 proportional units long. So, in order to keep the measurements in proportion, we need to set up the following proportion: $\frac{XY}{YZ} = \frac{1}{\sqrt{3}}$. Now substitute the known measurement of YZ from the figure, which is 5 in this problem.

$\frac{XY}{YZ} = \frac{1}{\sqrt{3}}$

$\frac{XY}{5} = \frac{1}{\sqrt{3}}$

$\left(\frac{XY}{5} \times 5\right) = \left(\frac{1}{\sqrt{3}} \times 5\right)$

$XY = \frac{5}{\sqrt{3}}$

64) The correct answer is C. Write out the formula: (length × 2) + (width × 2). Then substitute the values.
(17 × 2) + (4 × 2) = 34 + 8 = 42

65) The correct answer is C. A radian is the measurement of an angle at the center of a circle which is subtended by an arc that is equal in length to the radius of the circle. We need to use the formula to calculate the length of the arc: s = r θ. Substitute values to solve the problem.
radius (r) = 4
radians (θ) = π/4
s = r θ
s = 4 × π/4
s = π

66) The correct answer is C. Substitute the values from the problem.
cone volume = [height × radius² × π] ÷ 3
cone volume = [9 × 4² × π] ÷ 3
cone volume = [9 × 16 × π] ÷ 3
cone volume = 144π ÷ 3
cone volume = 48π

67) The correct answer is B. First, find the midpoint of the x coordinates for (−4, 2) and (8,−6).
midpoint $x = (x_1 + x_2) \div 2$

midpoint $x = (-4 + 8) \div 2$
midpoint $x = 4 \div 2$
midpoint $x = 2$
Then find the midpoint of the y coordinates for (−4, **2**) and (8,**−6**).
midpoint $y = (y_1 + y_2) \div 2$
midpoint $y = (2 + -6) \div 2$
midpoint $y = -4 \div 2$
midpoint $y = -2$
So, the midpoint is (2, −2)

68) The correct answer is D. Substitute the values provided (2, 3) and (6, 7) into the formula.
$d = \sqrt{(x_2 - x_1)^2 + (y_2 - y_1)^2}$
$d = \sqrt{(6 - 2)^2 + (7 - 3)^2}$
$d = \sqrt{4^2 + 4^2}$
$d = \sqrt{16 + 16}$
$d = \sqrt{32}$

69) The correct answer is A. Substitute the values into the slope-intercept formula.
$y = mx + b$
$315 = m5 + 15$
$315 - 15 = m5 + 15 - 15$
$300 = m5$
$300 \div 5 = m5 \div 5$
$60 = m$

70) The correct answer is A. The x intercept is the point at which a line crosses the x axis of a graph. In order for the line to cross the x axis, y must be equal to zero at that particular point of the graph. On the other hand, the y intercept is the point at which the line crosses the y axis. So, in order for the line to cross the y axis, x must be equal to zero at that particular point of the graph. First, substitute 0 for y in order to find the x intercept.
$x^2 + 2y^2 = 144$
$x^2 + (2 \times 0) = 144$
$x^2 + 0 = 144$
$x^2 = 144$
$x = 12$
Then substitute 0 for x in order to find the y intercept.
$x^2 + 2y^2 = 144$
$(0 \times 0) + 2y^2 = 144$
$0 + 2y^2 = 144$
$2y^2 \div 2 = 144 \div 2$
$y^2 = 72$
$y = \sqrt{72}$
So, the y intercept is (0, $\sqrt{72}$) and the x intercept is (12, 0).

71) The correct answer is D. You will recall from the formulas that sin A = cos (90° − A). If sin B = cos A, as in this problem, then B = 90° − A and A + B = 90°

72) The correct answer is C. Since the greatest possible value of cosine is 1, cos 2x must be less than or equal to 1. So, the greatest possible value of cos 2x is represented by the following formula: cos 2x = 1. Now, multiply each side of the equation by 4 in order to get 4 × cos 2x.
cos 2x = 1
4 × cos 2x = 1 × 4
4 × cos 2x = 4
So, the greatest possible value is 4.

73) The correct answer is A. $\sin^2$ of any angle is always equal to $1 - \cos^2$ of that angle.

74) The correct answer is C. From the trig formulas, we know that $\cos A = \frac{y}{z}$ (adjacent over hypotenuse) and $\tan A = \frac{x}{y}$ (opposite over adjacent). For our problem cos A = $\frac{9}{15}$ so the adjacent side is y = 9 and the hypotenuse z = 15. tan A = $\frac{12}{9}$ so the opposite side x = 12 and again y = 9. Now substitute the values for sine. For our problem, the opposite side x = 12 and the hypotenuse z = 15.
$\sin A = \frac{x}{z} = \frac{opposite}{hypotenuse}$, so for our question, $\sin A = \frac{12}{15}$.

75) The correct answer is C. Perform the operation inside the absolute value signs: −15 + 4 = −11. You find the absolute value by making the result of the operation inside the | | signs a positive number. So the absolute value of −11 = 11. |15| − |4| = 11, so it is the correct answer.

76) The correct answer is A. In this question, we need to calculate the slope of line A. Using the slope formula, we can calculate the slope of line A as follows: $\frac{y_2 - y_1}{x_2 - x_1} = \frac{3 - -5}{6 - 2} = \frac{8}{4} = 2$. The slopes of perpendicular lines are negative reciprocals of each other. So, to get the reciprocal for line B, you need to invert the integer to make a fraction: So, 2 becomes ½. You then need to make this a negative number, so ½ becomes − ½ . Line B has a y intercept of 0 because the facts of the question state that line B passes through the origin (0, 0). Using the slope intercept formula for line B with a slope of − ½ and a y intercept of 0, we get our answer: $y = -\frac{1}{2}x + 0$

77) The correct answer is A. The domain of a function is all possible x values for the function. On the other hand, the range is all of the possible y values for the function. Here, we need to find the range, so we are looking at y values. y will sometimes be positive and sometimes be negative, depending on the value of x. So, y can be any real number.

78) The correct answer is B. When the parentheticals are multiplied together, the result will be positive when x is a large positive number or when x is a large negative number. We then need to multiply by another positive number to get a positive output. So, the leading coefficient has to be positive if the end behavior is always positive.

79) The correct answer is B. The function $(x) = ax^2 + bx + c$ is graphed as a parabola. The other equation is graphed as a non-vertical line that intersects the parabola at two points.

80) The correct answer is A. Point A is (4, −6) at the start. The line is going to be shifted 2 units to the right (so we need to add 2 to the x coordinate) and 3 units down (so we need to subtract 3 from the y coordinate). So, the new position of point A will be (6, −9).

Solutions and Explanations for Practice Test Set 2 – Questions 81 to 160

81) The correct answer is B. In our problem, if $s\%$ have been absent, then $100\% - s\%$ have not been absent. In other words, since a percentage is any given number out of 100%, the percentage of students who have not been absent is: $(100\% - s\%)$. This equation is then multiplied by the total number of students (x) in order to determine the answer: $(100\% - s\%) \times x$

82) The correct answer is B. Line up the numbers by the comma, and remember to carry the 1's:

<pre>
  1 1 1
  1,594
+23,786
 25,380
</pre>

83) The correct answer is A. Multiply 25 times $15 to get the answer: $25 \times \$15 = \$375$

84) The correct answer is A. For problems with decimals, line the figures up in a column and add zeroes to fill in the column as shown below:

0.3500
0.0350
0.0530
0.3035

If you still struggle with decimals, you can remove the decimal points and the zeroes before the other integers in order to see the answer more clearly.

0.3500
0.0350
0.0530
0.3035

When we have removed the zeroes in front of the other numbers, we can clearly see that the largest number is 0.35.

85) The correct answer is C. Each panel is 8 feet 6 inches long, and he needs 11 panels to cover the entire side of the field. So, we need to multiply 8 feet 6 inches by 11, and then simplify the result. Step 1: 8 feet × 11 = 88 feet; Step 2: 6 inches × 11 = 66 inches; Step 3: There are 12 inches in a foot, so we need to determine how many feet and inches there are in 66 inches: 66 inches ÷ 12 = 5 feet 6 inches; Step 4: Now add the two results together. 88 feet + 5 feet, 6 inches = 93 feet 6 inches.

86) The correct answer is A. This problem is like question 84 above, except here we need to find a missing value. Remember to put in zeroes and line up the decimal points when you compare the numbers.

0.0007
A. 0.0012
B. 0.0006
C. 0.0022
D. 0.0220
0.0021

Answer choice B is less than 0.0007, and answer choices C and D are greater than 0.0021. Answer choice A (0.0012) is between 0.0007 and 0.0021, so it is the correct answer.

87) The correct answer is A. The equation is: F = $500P + $3,700. We are told that the total funds are $40,000 so put that in the equation to solve the problem.
$40,000 = $500P + $3,700
$40,000 − $3,700 = $500P
$36,300 = $500P
$36,300 ÷ 500 = $500 ÷ 500P
$36,300 ÷ 500 = 72.6
Since a fraction of a project cannot be undertaken, the greatest number of projects is 72.

88) The correct answer is D. To answer this type of question, you need these principles: (a) Positive numbers are greater than negative numbers; (b) When two fractions have the same numerator, the fraction with the smaller number in the denominator is the larger fraction. Accordingly, 1 is greater than $1/5$; $1/5$ is greater than $1/7$, and $1/7$ is greater than $-1/3$.

89) The correct answer is A. The problem tells us that the morning flight had 52 passengers more than the evening flight, and there were 540 passengers in total on the two flights that day. Step 1: First of all, we need to deduct the difference from the total: 540 − 52 = 488; In other words, there were 488 passengers on both flights combined, plus the 52 additional passengers on the morning flight. Step 2: Now divide this result by 2 to allocate an amount of passengers to each flight: 488 ÷ 2 = 244 passengers on the evening flight. (Had the question asked you for the amount of passengers on the morning flight, you would have had to add back the amount of additional passengers to find the total amount of passengers for the morning flight: 244 + 52 = 296 passengers on the morning flight)

90) The correct answer is C. Divide and then round up: 82 people in total ÷ 15 people served per container = 5.467 containers. We need to round up to 6 since we cannot purchase a fractional part of a container.

91) The correct answer is D. The question is asking us about a time duration of 6 minutes, so we need to calculate the amount of seconds in 6 minutes: 6 minutes × 60 seconds per minute = 360 seconds in total. Then divide the total time by the amount of time needed to make one journey: 360 seconds ÷ 45 seconds per journey = 8 journeys. Finally, multiply the number of journeys by the amount of inches per journey in order to get the total inches: 10.5 inches for 1 journey × 8 journeys = 84 inches in total

92) The correct answer is D. First of all, we need to find a common denominator for the fractions in the inequality, as well as for the fractions in the answer choices. In order to complete the problem quickly, you should not try to find the lowest common denominator, but just find any common denominator. We can do

this by expressing all of the numbers with a denominator of 90, since 9 is the largest denominator in the inequality and 10 is the largest denominator in the answer choices.

$2/3 \times 30/30 = 60/90$
$7/9 \times 10/10 = 70/90$

Then, express the original inequality in terms of the common denominator: $60/90 < ? < 70/90$
Next, convert the answer choices to the common denominator.
A) $1/3 \times 30/30 = 30/90$
B) $1/5 \times 18/18 = 18/90$
C) $2/6 \times 15/15 = 30/90$
D) $7/10 \times 9/9 = 63/90$

Finally, compare the results to find the answer. By comparing the numerators, we can see that $63/90$ lies between $60/90$ and $70/90$. So, D is the correct answer because $60/90 < 63/90 < 70/90$.

93) The correct answer is D. We have the following numbers in our problem:
0.0012
0.0253
0.2135
0.3152

If you still do not feel confident with decimals, remember that you can remove the decimal point and the zeroes after the decimal but before the other integers in order to see the answer more clearly.

   12
  253
 2135
 3152

94) The correct answer is C. If $\frac{x}{24}$ is between 8 and 9, x will need to be between 192 and 216, since $\frac{192}{24}$ =192 ÷ 24 = 8 and $\frac{216}{24}$ = 216 ÷ 24 = 9. 200 is the only number from the answer choices that is greater than 192 and less than 216.

95) The correct answer is C. Work backwards based on the facts given. There are 18 students left at the end after one-fourth of them left for the principal's office. So, set up an equation for this:
18 + $1/4$T = T
18 + $1/4$T – $1/4$T = T – $1/4$T
18 = $3/4$T
18 × 4 = $3/4$T × 4
72 = 3T
72 ÷ 3 = 3T ÷ 3
24 = T

So, before the group of pupils left to see the principal, there were 24 students in the class. We know that one-fifth of the students left at the beginning to go to singing lessons, so we need to set up an equation for this:

$24 + \frac{1}{5}T = T$

$24 + \frac{1}{5}T - \frac{1}{5}T = T - \frac{1}{5}T$

$24 = \frac{4}{5}T$

$24 \times 5 = \frac{4}{5}T \times 5$

$120 = 4T$

$120 \div 4 = 4T \div 4$

$30 = T$

96) The correct answer is B. At the beginning of January, there are 300 students, but 5% of the students leave during the month, so we have 95% left at the end of the month: 300 × 95% = 285. Then 15 students join on the last day of the month, so we add that back in to get to the total at the end of January: 285 + 15 = 300. If this pattern continues, there will always be 300 students in the academy at the end of any month.

97) The correct answer is D. Calculate the discount: $120 × 12.5% = $15 discount. Then subtract the discount from the original price to determine the sales price: $120 – $15 = $105

98) The correct answer is C. Divide by the fractional hour in order to determine the speed for an entire hour: 38.4 miles ÷ $\frac{4}{5}$ of an hour = 38.4 × $\frac{5}{4}$ = (38 × 5) ÷ 4 = 47.5 mph. We round this up to 48 mph.

99) The correct answer is A. The ratio of defective chips to functioning chips is 1 to 20. So, the defective chips form one group and the functioning chips form another group. Therefore, the total data set can be divided into groups of 21. Accordingly, $\frac{1}{21}$ of the chips will be defective. The factory produced 11,235 chips last week, so we calculate as follows: 11,235 × $\frac{1}{21}$ = 535

100) The correct answer is B. The total amount available is $55,000, so we can substitute this for C in the equation provided in order to calculate R number of residents:

C = $750R + $2,550

$55,000 = $750R + $2,550

$55,000 – $2,550 = $750R + $2,550 – $2,550

$55,000 – $2,550 = $750R

$52,450 = $750R

$52,450 ÷ $750 = $750R ÷ $750

$52,450 ÷ $750 = R

69.9 = R

It is not possible to accommodate a fractional part of one person, so we need to round down to 69 residents.

101) The correct answer is B. First, determine how many cheese and pepperoni pizzas were sold. Each triangle symbol represents 5 pizzas. Therefore, 15 cheese pizzas were sold: 3 symbols on the pictograph × 5 pizzas per symbol = 15 cheese pizzas. We also know that 10 pepperoni pizzas were sold: 2 symbols on the pictograph × 5 pizzas per symbol = 10 pepperoni pizzas. Then determine the value of these two types of pizzas based on the prices stated in the problem: (15 cheese pizzas × $10 each) + (10 pepperoni pizzas × $12 each) = $150 + $120 = $270. The remaining amount is allocable to the vegetable

pizzas: Total sales of $310 − $270 = $40 worth of vegetable pizzas. Since each triangle represents 5 pizzas, 5 vegetable pizzas were sold. We calculate the price of the vegetable pizzas as follows:
$40 worth of vegetable pizzas ÷ 5 vegetable pizzas sold = $8 per vegetable pizza

102) The correct answer is D. This is a problem on determining the value that is missing from the calculation of a mean of a set of values. Remember that the mean is the same thing as the arithmetic average. In order to calculate the mean, you simply add up the values of all of the items in the set, and then divide by the number of items in the set. Set up your equation to calculate the average, using x for the age of the 5th sibling:
$(2 + 5 + 7 + 12 + x) \div 5 = 8$
$(2 + 5 + 7 + 12 + x) \div 5 \times 5 = 8 \times 5$
$(2 + 5 + 7 + 12 + x) = 40$
$26 + x = 40$
$26 − 26 + x = 40 − 26$
$x = 14$

103) The correct answer is A. To solve problems like this one, it is usually best to write out the possible outcomes in a list. This will help you visualize the number of possible outcomes that make up the sample space. Then circle or highlight the events from the list to get your answer. In this case, we have two items, each of which has a variable outcome. There are 6 numbers on the black die and 6 numbers on the red die. Using multiplication, we can see that there are 36 possible combinations: 6 × 6 = 36
To check your answer, you can list the possibilities of the various combinations:

(1,1) (1,2) (1,3) (1,4) (1,5) (1,6)
(2,1) (2,2) (2,3) (2,4) (2,5) (2,6)
(3,1) (3,2) (3,3) (3,4) (3,5) (3,6)
(4,1) (4,2) (4,3) (4,4) (4,5) (4,6)
(5,1) (5,2) (5,3) **(5,4)** (5,5) (5,6)
(6,1) (6,2) (6,3) (6,4) (6,5) (6,6)

If the number on the left in each set of parentheses represents the black die and the number on the right represents the red die, we can see that there is one chance that Sam will roll a 4 on the red die and a 5 on the black die. The result is expressed as a fraction, with the event (chance of the desired outcome) in the numerator and the total sample space (total data set) in the denominator. So, the answer is $1/36$.

104) The correct answer is D. The median is the number that is halfway through the set. Our data set is: 19, 20, 20, 15, 21, 18, 20, 23, 22, 15, 12, 23, 9, 18, 17. So first, put the numbers in ascending order:
9, 12, 15, 15, 17, 18, 18, 19, 20, 20, 20, 21, 22, 23, 23. We have 15 numbers, so the 8th number in the set is halfway and is therefore the median: 9, 12, 15, 15, 17, 18, 18, **19**, 20, 20, 20, 21, 22, 23, 23

105) The correct answer is C. Find the percentage for the patients that have not survived: 100% − 48% = 52%. Then multiply that percentage by the total for this category: 231,000 × 52% = 120,120

106) The correct answer is A. Try to find the month where all three lines look the highest. This appears to be May, November or December. Add up the three amounts for these months to check which one is highest. May 1,400,000 + 600,000 + 400,000 = 2,400,000; November: 1,300,000 + 1,000,000 + 900,00 = 3,200,000; December: 1,500,000 + 1,000,000 + 1,000,000 = 3,500,000. So, December is the highest in total for all three companies.

107) The correct answer is D. This type of bar graph is called a histogram. For questions on histograms, you need to add up the bars of the same color to find the total amounts for each group. Don't worry if you are not sure of the exact amounts for each bar. Just try to get as close as possible. Here, we add up for each country as follows: Cobb County: 2.8 + 2.1 + 1.2 + 0.8 = 6.9; Dawson County: 3.5 + 1.1 + 0.9 + 2.3 = 7.8; Emery County: 2.5 + 1.8 + 1 + 0.9 = 6.2. So, 7.8 inches is the greatest amount for any one county in total.

108) The correct answer is D. You need to determine the amount of possible outcomes at the start of the day first of all. The owner has 10 brown teddy bears, 8 white teddy bears, 4 black teddy bears, and 2 pink teddy bears when she opens the attraction at the start of the day. So, at the start of the day, she has 24 teddy bears: 10 + 8 + 4 + 2 = 24. Then you need to reduce this amount by the quantity of items that have been removed. The problem tells us that she has given out a brown teddy bear, so there are 23 teddy bears left in the sample space: 24 − 1 = 23. The event is the chance of the selection of a pink teddy bear. We know that there are two pink teddy bears left after the first prize winner receives his or her prize. Finally, we need to put the event (the number representing the chance of the desired outcome) in the numerator and the number of possible remaining combinations (the sample space) in the denominator. So the answer is $2/23$.

109) The correct answer is B. Our data set is: 2.5, 9.4, 3.1, 1.7, 3.2, 8.2, 4.5, 6.4, 7.8. First, put the numbers in ascending order: 1.7, 2.5, 3.1, 3.2, 4.5, 6.4, 7.8, 8.2, 9.4. The median is the number in the middle of the set: 1.7, 2.5, 3.1, 3.2, **4.5**, 6.4, 7.8, 8.2, 9.4

110) The correct answer is D. The ratio of bags of apples to bags of oranges is 2 to 3, so for every two bags of apples, there are three bags of oranges. First, take the total amount of bags of apples and divide by 2: 44 ÷ 2 = 22. Then multiply this by 3 to determine how many bags of oranges there are in the store: 22 × 3 = 66.

111) The correct answer is B.
Factor: $x^2 − 5x + 6 \leq 0$
$(x − 2)(x − 3) \leq 0$
Then solve each parenthetical for zero:
$(x − 2) = 0$
$2 − 2 = 0$
$x = 2$

$(x − 3) = 0$
$3 − 3 = 0$
$x = 3$
So, $2 \leq x \leq 3$

Now check. Use 1 to check to $2 \leq x$, which is the same as $x \geq 2$. Since 1 is not actually greater than or equal to 2, our proof for this should be incorrect.
$x^2 − 5x + 6 \leq 0$
$1^2 − (5 × 1) + 6 \leq 0$
$1 − 5 + 6 \leq 0$
$−4 + 6 \leq 0$
$2 \leq 0$ INCORRECT

Use 2.5 to check for x ≤ 3. Since 2.5 really is less than 3, our proof should be correct.
$x^2 - 5x + 6 \leq 0$
$2.5^2 - (5 \times 2.5) + 6 \leq 0$
$6.25 - 12.5 + 6 \leq 0$
$-0.25 \leq 0$ CORRECT

Therefore, we have checked that $2 \leq x \leq 3$

112) The correct answer is B. Substitute 12 for the value of x. Then simplify and solve.
$x^2 + xy - y = 254$
$12^2 + 12y - y = 254$
$144 + 12y - y = 254$
$144 - 144 + 12y - y = 254 - 144$
$12y - y = 110$
$11y = 110$
$11y \div 11 = 110 \div 11$
$y = 10$

113) The correct answer is A.
FIRST: $(\mathbf{3x} + y)(\mathbf{x} - 5y) = 3x \times x = 3x^2$
OUTSIDE: $(\mathbf{3x} + y)(x - \mathbf{5y}) = 3x \times -5y = -15xy$
INSIDE: $(3x + \mathbf{y})(\mathbf{x} - 5y) = y \times x = xy$
LAST: $(3x + \mathbf{y})(x - \mathbf{5y}) = y \times -5y = -5y^2$
Then add all of the above once you have completed FOIL: $3x^2 - 15xy + xy - 5y^2 = 3x^2 - 14xy - 5y^2$

114) The correct answer is A. The factors of 9 are: 1 × 9 = 9; **3** × 3 = 9. The factors of 3 are: 1 × **3** = 3. So, put the integer for the common factor outside the parentheses first: $9x^3 - 3x = 3(3x^3 - x)$
Then determine if there are any common variables for the terms that remain in the parentheses.
For $(3x^2 - x)$ the terms $3x^2$ and x have the variable x in common. So, now factor out x to solve:
$3(3x^3 - x) = 3x(3x^2 - 1)$

115) The correct answer is C. This is a square, so to find the length of one side, we divide the perimeter by four: 24 ÷ 4 = 6. Now we use the Pythagorean theorem to find the length of line segment AB. In this case AB is the hypotenuse. The hypotenuse length is the square root of $6^2 + 6^2$.
$\sqrt{6^2 + 6^2} = \sqrt{36 + 36} = \sqrt{72}$

116) The correct answer is A. As y increases by 5, x decreases by 5. So, the slope is –1. The line includes point (20, 15), which is the fifth point from the left.

117) The correct answer is B. Add the numbers in front of the radical signs to solve. If there is no number before the radical, then put in the number 1 because then the radical will count only 1 time when you add.
$\sqrt{15} + 3\sqrt{15} = 1\sqrt{15} + 3\sqrt{15} = (1 + 3)\sqrt{15} = 4\sqrt{15}$

118) The correct answer is C. In order to multiply two square roots, multiply the numbers inside the radical signs: $\sqrt{5} \times \sqrt{3} = \sqrt{5 \times 3} = \sqrt{15}$

119) The correct answer is B. Find the lowest common denominator. Then add the numerators together as shown: $\frac{x}{5} + \frac{y}{2} = \left(\frac{x}{5} \times \frac{2}{2}\right) + \left(\frac{y}{2} \times \frac{5}{5}\right) = \frac{2x}{10} + \frac{5y}{10} = \frac{2x + 5y}{10}$

120) The correct answer is D. The slope intercept formula is: $y = mx + b$. Remember that m is the slope and b is the y intercept. You will also need the slope formula: $m = \frac{y_2 - y_1}{x_2 - x_1}$

We are given the slope, as well as point (4,5), so first we need to put those points into the slope formula. We are doing this in order to solve for b, which is not provided in the facts of the problem.

$$\frac{y_2 - y_1}{x_2 - x_1} = -\frac{3}{5}$$

$$\frac{5 - y_1}{4 - x_1} = -\frac{3}{5}$$

Then eliminate the denominator.

$$(4 - x_1)\frac{5 - y_1}{4 - x_1} = -\frac{3}{5}(4 - x_1)$$

$$5 - y_1 = -\frac{3}{5}(4 - x_1)$$

Now put in 0 for x_1 in the slope formula in order to find b, which is the y intercept (the point at which the line crosses the y axis).

$$5 - y_1 = -\frac{3}{5}(4 - x_1)$$

$$5 - y_1 = -\frac{3}{5}(4 - 0)$$

$$5 - y_1 = -\frac{3 \times 4}{5}$$

$$5 - y_1 = -\frac{12}{5}$$

$$5 - 5 - y_1 = -\frac{12}{5} - 5$$

$$-y_1 = -\frac{12}{5} - 5$$

$$-y_1 \times -1 = \left(-\frac{12}{5} - 5\right) \times -1$$

$$y_1 = \frac{12}{5} + 5$$

$$y_1 = \frac{12}{5} + \left(5 \times \frac{5}{5}\right)$$

$$y_1 = \frac{12}{5} + \frac{25}{5}$$

$$y_1 = \frac{37}{5}$$

Remember that the *y* intercept (known in the slope-intercept formula as the variable *b*) exists when *x* is equal to 0. We have put in the value of 0 for *x* in the equation above, so $b = \frac{37}{5}$. Now put the value for *b* into the slope intercept formula.

$$y = mx + b$$
$$y = -\frac{3}{5}x + \frac{37}{5}$$

121) The correct answer is D. Factor and cancel out if possible. Then multiply.

$$\frac{x^2 + 5x + 4}{x^2 + 6x + 5} \times \frac{16}{x + 5} =$$

$$\frac{(x + 1)(x + 4)}{(x + 1)(x + 5)} \times \frac{16}{x + 5} =$$

$$\frac{\cancel{(x + 1)}(x + 4)}{\cancel{(x + 1)}(x + 5)} \times \frac{16}{x + 5} =$$

$$\frac{(x + 4)}{(x + 5)} \times \frac{16}{x + 5} =$$

$$\frac{(x + 4) \times 16}{(x + 5)(x + 5)} =$$

$$\frac{16x + 64}{x^2 + 10x + 25}$$

122) The correct answer is D. When dividing fractions, you need to invert the second fraction and then multiply the two fractions together.

$$\frac{8x - 8}{x} \div \frac{3x - 3}{6x^2} = \frac{8x - 8}{x} \times \frac{6x^2}{3x - 3}$$

Then look at the numerator and denominator from the result of the previous step to see if you can factor and cancel out.

$$\frac{8x - 8}{x} \times \frac{6x^2}{3x - 3} =$$

$$\frac{8(x - 1)}{x} \times \frac{6x^2}{3(x - 1)} =$$

$$\frac{8\cancel{(x - 1)}}{x} \times \frac{6x^2}{3\cancel{(x - 1)}} =$$

$$\frac{8 \times 6x^2}{x \times 3} =$$

$$\frac{8 \times (2 \times 3 \times x \times x)}{x \times 3} =$$

$$\frac{8 \times (2 \times \cancel{3} \times \cancel{x} \times x)}{\cancel{x} \times \cancel{3}} = 16x$$

123) The correct answer is C. Any non-zero number to the power of zero is equal to 1.

124) The correct answer is B. $4^{11} \times 4^8 = 4^{(11+8)} = 4^{19}$

125) The correct answer is C. Perform the operation on the radicals and then simplify.
$\sqrt{8x^4} \cdot \sqrt{32x^6} = \sqrt{8x^4 \times 32x^6} = \sqrt{256x^{10}} = \sqrt{16 \times 16 \times x^5 \times x^5} = 16x^5$

126) The correct answer is D. Use the formula: volume = base × width × height = 20 × 15 × 25 = 7500

127) The correct answer is D. Substitute the values into the formula in order to find the solution.
$\sqrt{A^2 + B^2} = C$
$\sqrt{5^2 + 12^2} = C$
$\sqrt{25 + 144} = C$
$\sqrt{169} = C$
13 cm

128) The correct answer is C. Corresponding angles are equal in measure. So, for example, angles r and u are equal, and angles s and v are equal. Opposite angles will be equal when bisected by two parallel lines. Angles s and t are opposite, and angles u and w are also opposite. So, ∠r, ∠u, and ∠w are equal.

129) The correct answer is A. We need to express $\frac{3}{18}\pi$ in degrees. We know from the radian formulas that π × radian = 180°. So, we can substitute 180° for π in our equation.
$\frac{3}{18}\pi = \frac{3}{18} 180° = 30°$

130) The correct answer is C. Essentially a rectangle is missing at the upper left-hand corner of the figure. We would need to know both the length and width of the "missing" rectangle in order to calculate the area of our figure. So, we need to know both X and Y in order to solve the problem.

131) The correct answer is D. The prism has 5 sides, so we need to calculate the surface area of each one. The rectangle at the bottom of the prism that lies along points, A, B, and D measures 3.5 units (side AB) by 5 units (side BD), so the surface area of the bottom rectangle is: Length × Width = 3.5 × 5 = 17.5 Then calculate the area of the rectangle at the back of the triangle, lying along points A and C. This rectangle measures 4 units (side AC) by 5 units (the side that is parallel to side BD). So, the area of this side is: Length × Width = 4 × 5 = 20. Next we need to find the length of the hypotenuse (side CB). Since AB is 3.5 units and AC is 4 units, we can use the Pythagorean theorem as follows:
$\sqrt{3.5^2 + 4^2} = \sqrt{12.25 + 16} = \sqrt{28.25} \approx 5.3$. We can then calculate the surface area of the sloping rectangle that lies along the hypotenuse (along points C, B and D) as: Length × Width = 5.3 × 5 = 26.5

Next, we need to calculate the surface area of the two triangles on each end of the prism. The formula for the area of a triangle is $bH \div 2$, so substituting the values we get: $(3.5 \times 4) \div 2 = 7$
Finally, add the area of all five sides together to get the surface area for the entire prism:
$17.5 + 20 + 26.5 + 7 + 7 = 78$

132) The correct answer is C. Volume of cylinder = $\pi R^2 h = \pi \times$ radius$^2 \times$ height. In our problem, R = 5 and h = 14. $\pi R^2 h = \pi 5^2 \times 14 = 25\pi \times 14 = 350\pi$

133) The correct answer is D. The formula for the area of a circle is: $\pi \times R^2$. First, we need to calculate the area of the larger circle: $\pi \times 2.4^2 = 5.76\pi$. Then calculate the area of the smaller inner circle: $\pi \times 1^2 = \pi$. We need to find the difference between half of each circle, so divide the area of each circle by 2 and then subtract:
$$(5.76\pi \div 2) - (\pi \div 2) = \frac{5.76\pi}{2} - \frac{\pi}{2} = \frac{4.76\pi}{2} = 2.38\pi$$

134) The correct answer is D. First, calculate the area of the central rectangle. The area of a rectangle is length times height: $8 \times 3 = 24$. Then we use the Pythagorean theorem to work out that the base of each triangle is 4.
$5 = \sqrt{3^2 + base^2}$
$5^2 = 3^2 + base^2$
$25 = 9 + base^2$
$25 - 9 = 9 - 9 + base^2$
$16 = base^2$
$4 = base$
Then calculate the area of each of the triangles on each side of the central rectangle. The area of a triangle is base times height divided by 2: $(4 \times 3) \div 2 = 6$. So, the total area is the area of the main rectangle plus the area of each of the two triangles: $24 + 6 + 6 = 36$

135) The correct answer is D. The area of a circle is π times the radius squared. Therefore, the area of circle A is: $3^2\pi = 9\pi$. Since the circles are internally tangent, the radius of circle B is calculated by taking the radius of circle A times 2. In other words, the diameter of circle A is the radius of circle B. Therefore, the radius of circle B is $3 \times 2 = 6$ and the area of circle B is $6^2\pi = 36\pi$. To calculate the remaining area of circle B, we subtract as follows: $36\pi - 9\pi = 27\pi$

136) The correct answer is C. Circumference = $\pi \times 2 \times$ radius. In our question, the radius is 12, so the circumference is 24π.

137) The correct answer is B. The area of circle M is $8^2\pi = 64\pi$. The area of circle M is 39π greater than the area of circle N, so subtract to find the area of circle N: $64\pi - 39\pi = 25\pi$. The area of circle N is calculated as follows: $5^2\pi = 25\pi$. So the radius of circle N is 5.

138) The correct answer is A. Remember that the y intercept is where the line crosses the y axis, so x = 0 for the y intercept. Begin by substituting 0 for x.
y = x + 14
y = 0 + 14
y = 14

Therefore, the coordinates (0, 14) represent the y intercept.

On the other hand, the x intercept exists where the line crosses the x axis, so y = 0 for the x intercept. Now substitute 0 for y.
y = x + 14
0 = x + 14
0 − 14 = x + 14 − 14
−14 = x
So, the coordinates (−14, 0) represent the x intercept.

139) The correct answer is A. Our points are (5, 2) and (7, 4), so substitute the values into the midpoint formula.
$(x_1 + x_2) \div 2$, $(y_1 + y_2) \div 2$
(5 + 7) ÷ 2 = midpoint x, (2 + 4) ÷ 2 = midpoint y
12 ÷ 2 = midpoint x, 6 ÷ 2 = midpoint y
6 = midpoint x, 3 = midpoint y

140) The correct answer is D. $x = \log_y Z$ is the same as $y^x = Z$, so $6^4 = 1296$ is the same as $4 = \log_6 1296$

141) The correct answer is C. y is positive even when x is negative, so we know that x is squared. We have an upward pointing parabola, so we know that the leading coefficient must be positive. Looking at the value for x = 0, we can see that the output is for y is 5, so the correct function is $f(x) = x^2 + 5$.

142) The correct answer is C. Put the value provided for x into the function to solve. $f_1(2) = 5^2 = 25$

143) The correct answer is B. The principle is that $x^{-b} = \dfrac{1}{x^b}$. Therefore, $(-4)^{-3} = \dfrac{1}{-4^3} = -\dfrac{1}{64}$

144) The correct answer is B. Use the Pythagorean theorem:
$AB^2 + BC^2 = AC^2$
$AB^2 + 8^2 = 10^2$
$AB^2 + 64 = 100$
$AB^2 + 64 − 64 = 100 − 64$
$AB^2 = 36$
AB = 6
AB is the adjacent side and BC is the opposite side. Tangent is opposite over adjacent, so the tangent of angle A equals BC divided by AB: 8 ÷ 6 = $^4/_3$

145) The correct answer is C. XY is the opposite side and XZ is the hypotenuse. Sine is "opposite over hypotenuse", so the sine of angle Z is calculated by dividing XY by XZ.
sin z = $^{XY}/_{XZ}$ = $^{opposite}/_{hypotenuse}$
sin z = $^{XY}/_{12}$
Since angle Z is 30 degrees, and the facts state that sin 30° = 0.5, we can substitute values as follows:
sin z = $^{XY}/_{12}$
0.5 = $^{XY}/_{12}$
0.5 × 12 = $^{XY}/_{12}$ × 12

0.5 × 12 = XY
6 = XY

146) The correct answer is D. This question tests your recall of the trigonometric formulas. In this problem, $\cos A = b/c$, so b is the adjacent side and c is the hypotenuse. $\sin A = a/c$, so a is the opposite side. Tangent equals "opposite over adjacent", so $\tan A = a/b$.

147) The correct answer is A. $\tan A = \sin A \div \cos A$, so substitute the values into the formula.
tan A = sin A ÷ cos A
tan A = 0.95106 ÷ 0.30902
tan A = 3.07767

148) The correct answer is C. The plumber is going to earn $4,000 for the month. He charges a set fee of $100 per job, and he will do 5 jobs, so we can calculate the total set fees first: $100 set fee per job × 5 jobs = $500 total set fees. Then deduct the set fees from the total for the month in order to determine the total for the hourly pay: $4,000 – $500 = $3,500. He earns $25 per hour, so divide the hourly rate into the total hourly pay in order to determine the number of hours he will work: $3,500 total hourly pay ÷ $25 per hour = 140 hours to work

149) The correct answer is D. $(2 + \sqrt{6})^2 = (2 + \sqrt{6})(2 + \sqrt{6}) =$
$(2 \times 2) + (2 \times \sqrt{6}) + (2 \times \sqrt{6}) + (\sqrt{6} \times \sqrt{6}) = 4 + 4\sqrt{6} + 6 = 10 + 4\sqrt{6}$

150) The correct answer is B. The trig formulas state that tan A × cos A = sin A. In this question, the cosine of angle Z is 0.78801075 and the tangent of angle Z is 0.7812856, so put the values into the formula to solve.
tan A × cos A = sin A
0.7812856 × 0.78801075 = sin A
0.61566148 = sin A

151) The correct answer is D. Using the slope formula, we can calculate the slope of the line as follows: $\frac{y_2 - y_1}{x_2 - x_1} = \frac{8 - 4}{6 - 0} = \frac{4}{6} = \frac{2}{3}$. So, variable m for the slope needs to be $\frac{2}{3}$ for our equation. The line crosses the y axis at point (0, 4), so the y intercept is 4. Finally, put these values into the equation of a line to solve: $y = \frac{2}{3}x + 4$

152) The correct answer is B. First, you need to convert the logarithmic function into an exponential equation. To convert a logarithmic function to an exponent, the number after the equals sign (4 in this problem) becomes the exponent. The small subscript number after "log" (3 in this problem) becomes the base number. So, the exponential equation for $\log_3(x + 2) = 4$ is $3^4 = x + 2$. Then find the result for the exponent: $3^4 = 3 \times 3 \times 3 \times 3 = 81$. Substituting 81 on the left side of the equation, we get $81 = x + 2$. Therefore, $x = 79$.

153) The correct answer is C. Remember that the domain of a function is all possible x values for the function. You need to avoid any mathematical operations that do not have real number solutions, such as

dividing by a zero or finding the square root of a negative number. To avoid dividing by a zero in our problem, $x \neq 2$. Therefore, the domain is all real numbers except 2.

154) The correct answer is C. Before we shift the figure, point C has the coordinates (–4, –3). We are moving the figure 7 units to the right, thereby adding 7 to the x coordinate, and 6 units down, thereby subtracting 6 from the y coordinate. –4 + 7 = 3 and –3 – 6 = –9, so the new coordinates are: (3, –9).

155) The correct answer is D. The range of a function is all possible y values or "outputs" for the function. $x^2 + 36$ will yield a result of 36 when $x = 0$. $x^2 + 36$ will result in a positive number greater than 36 for all other values of x. Therefore, the range will always be equal to or greater than 36.

156) The correct answer is C. The y intercept is where $x = 0$. So, we can substitute 0 in our equation to solve: ($12 + 2x$) ÷ (4 + x) = ($12 + $0) ÷ (4 + 0) = $12 ÷ 4 = 3.

157) The correct answer is D. If $x = 3$ and the value above the sigma sign is 6, you need to find the individual products for $x^2 + 5$ for $x = 3$, $x = 4$, $x = 5$, and $x = 6$.
For $x = 3$: $x^2 + 5 = 3^2 + 5 = 14$
For $x = 4$: $x^2 + 5 = 4^2 + 5 = 21$
For $x = 5$: $x^2 + 5 = 5^2 + 5 = 30$
For $x = 6$: $x^2 + 5 = 6^2 + 5 = 41$
Then you add these four products together to solve: 14 + 21 + 30 + 41 = 106

158) The correct answer is A. The total amount that Toby has to pay is represented by C. He is paying D dollars immediately, so we can determine the remaining amount that he owes by deducting his down payment from the total. So, the remaining amount owing is represented by the equation: C – D. We have to divide the remaining amount owing by the number of months (M) to get the monthly payment (P):
P = (C – D) ÷ M = $\frac{C-D}{M}$

159) The correct answer is B. Assign a variable for the age of each brother. Alex = A, Burt = B, and Zander = Z. Alex is twice as old as Burt, so A = 2B. Burt is one year older than three times the age of Zander, so B = 3Z + 1. Then substitute the value of B into the first equation.
A = 2B
A = 2(3Z + 1)
A = 6Z + 2
So, Alex is 2 years older than 6 times the age of Zander.

160) The correct answer is D. The original price of the sofa on Wednesday was x. On Thursday, the sofa was reduced by 10%, so the price on Thursday was 90% of x or $0.90x$. On Friday, the sofa was reduced by a further 15%, so the price on Friday was 85% of the price on Thursday, so we can multiply Thursday's price by 0.85 to get our answer: $(0.90)(0.85)x$

Solutions and Explanations for Practice Test Set 3 – Questions 161 to 240

161) The correct answer is C. The value of μ must be greater than $^{11}/_3$, which is equal to 3.6667. The answer 4.1 is the only option which meets this criterion.

162) The correct answer is C. Remember that the order of operations is PEMDAS: Parentheses, Exponents, Multiplication, Division, Addition, and Subtraction. There are no operations with parentheses, exponents, or multiplication. So, do the division first: 9 ÷ 3 = 3. Then replace this in the equation: 82 + 9 ÷ 3 − 5 = 82 + 3 − 5 = 80

163) The correct answer is A. This is another problem on the order of operations. There are no operations with parentheses or exponents, so do the multiplication first: 6 × 3 = 18. Then put this number in the equation: 52 + 6 × 3 − 48 = 52 + 18 − 48 = 22

164) The correct answer is C. In order to convert a fraction to a decimal, you must divide.

```
      .25
16)4.00
    3.2
    0.80
    0.80
       0
```

165) The correct answer is D. 30 percent in decimal form equals to 0.30. The phrase "of what number" shows that we have to divide: 90 ÷ 0.30 = 300. We can check this result as follows: 300 × 0.30 = 90.

166) The correct answer is A. Questions like this test your knowledge of mixed numbers. Mixed numbers are those that contain a whole number and a fraction. If the fraction on the first mixed number is greater than the fraction on the second mixed number, you can subtract the whole numbers and the fractions separately. Remember to use the lowest common denominator on the fractions. First, subtract whole numbers: 6 − 2 = 4

Then subtract fractions.
$3/4 - 1/2 =$
$3/4 - 2/4 =$
$1/4$

Now put them together for the result.
$4 \, 1/4$

167) The correct answer is D. Remember PEMDAS: Parentheses, Exponents, Multiplication, Division, Addition, and Subtraction. So, you must do the division and multiplication first, before adding or subtracting: 9 × 6 + 42 ÷ 6 = (9 × 6) + (42 ÷ 6). We know that 9 × 6 = 54 and 42 ÷ 6 = 7 so perform the operations and simplify: (9 × 6) + (42 ÷ 6) = 54 + 7 = 61.

168) The correct answer is B. To fnd the total amount contributed, you need to multiply.
31 × 12 = 372

169) The correct answer is D. When you are asked to divide fractions, remember that you need to invert the second fraction. Then you multiply this inverted fraction by the first fraction given in the problem. $4/3$ inverted is $3/4$. Then multiply the numerators and the denominators together to get the new fraction.

$$\frac{1}{8} \div \frac{4}{3} = \frac{1}{8} \times \frac{3}{4} = \frac{3}{32}$$

170) The correct answer is D. Convert the fractions in the mixed numbers to decimals.
$3/4 = 3 \div 4 = 0.75$
$1/5 = 1 \div 5 = 0.20$

Then represent the mixed numbers as decimal numbers.
Person 1: $14^{3}/_{4}$ = 14.75; Person 2: $20^{1}/_{5}$ = 20.20; Person 3: 36.35

Then add all three amounts together to find the total: 14.75 + 20.20 + 36.35 = 71.30

171) The correct answer is C. Remember that to represent a fraction as a decimal, you need to divide. So, you will need to do long division to determine the answer.

```
      .10
50)5.00
    5.00
       0
```

172) The correct answer is D. Be careful not to confuse remainders with decimals. The remainder is the whole number amount left over after you have used whole numbers to divide.

```
      66
9)600
    54
    60
    54
     6 – This is the remainder.
```

173) The correct answer is A. This question assesses your knowledge of mixed numbers. In this problem, the fraction on the second number is bigger than the fraction on the first number. So, we have to convert the mixed numbers to fractions first.

$$3\frac{1}{2} - 2\frac{3}{5} = \left[\left(3 \times \frac{2}{2}\right) + \frac{1}{2}\right] - \left[\left(2 \times \frac{5}{5}\right) + \frac{3}{5}\right] = \left[\frac{6}{2} + \frac{1}{2}\right] - \left[\frac{10}{5} + \frac{3}{5}\right] = \frac{7}{2} - \frac{13}{5} =$$

Then find the lowest common denominator.

$$\frac{7}{2} - \frac{13}{5} = \left(\frac{7}{2} \times \frac{5}{5}\right) - \left(\frac{13}{5} \times \frac{2}{2}\right) = \frac{35}{10} - \frac{26}{10} = \frac{9}{10}$$

174) The correct answer is B. Remember for division of fractions, you need to invert the second fraction and then multiply the fractions. When you multiply fractions, you multiply the numerators with each other for the new numerator, and the denominators with each other for the new denominator. For problems like this, deal with the parts of the equation in the parentheses first.

$$\frac{1}{6}+\left(\frac{1}{2}\div\frac{3}{8}\right)-\left(\frac{1}{3}\times\frac{3}{2}\right)=\frac{1}{6}+\left(\frac{1}{2}\times\frac{8}{3}\right)-\left(\frac{1}{3}\times\frac{3}{2}\right)=\frac{1}{6}+\frac{8}{6}-\frac{3}{6}$$

After you have done the operations on the parentheses, you can add and subtract as needed.

$$\frac{1}{6}+\frac{8}{6}-\frac{3}{6}=\frac{9}{6}-\frac{3}{6}=\frac{6}{6}=1$$

175) The correct answer is D. We know that Mary has already gotten 80% of the money. However, the question is asking how much money she still needs: 100% − 80% = 20% = .20
Now do the multiplication: 650 × .20 = 130

176) The correct answer is B. Your equation is: (A × $5) + (B × $8) = $60. They buy 4 of product A, so put that in the equation and solve it.
(A × $5) + (B × $8) = $60
(4 × $5) + (B × $8) = $60
$20 + (B × $8) = $60
(B × $8) = $40
B = 5

177) The correct answer is D. For practical problems like this, you must first determine the percentage that you need in order to solve the problem. Then, you must do long multiplication to determine how many games the team won. The question tells you the percentage of games the team lost, not won.
STEP 1: First of all, we have to calculate the percentage of games won. If the team lost 20 percent of the games, we know that the team won the remaining 80 percent.
STEP 2: Now do the long multiplication.
   50 games in total
× .80 percentage of games won (in decimal form)
 40.0 total games won

178) The correct answer is B. STEP 1: Look to see what information is common to both the question and to the information provided. Here we have the information that he can run 3 miles in 25 minutes. The question is asking how long it will take him to run 12 miles, so the commonality is miles. STEP 2: Next, you need to find out how many 3-mile increments there are in 12 miles: 12 ÷ 3 = 4. STEP 3: Then you need to determine the time required to travel the stated distance. Accordingly, we need to multiply the time for 3 miles by 4: 25 minutes × 4 = 100. So, 100 minutes are needed to run 12 miles. STEP 4: Finally, simplify into hours and minutes based on the fact that there are 60 minutes in one hour: 100 minutes = 1 hour 40 minutes.

179) The correct answer is A. STEP 1: First of all, we need to find out how many pieces of candy there are in total: 43 + 28 + 31 = 102 total pieces of candy. STEP 2: We need to divide the total amount of candy by the number of students in order to find out how much candy each student will get: 102 total pieces of candy ÷ 34 students = 3 pieces of candy per student.

180) The correct answer is D. The lowest temperature is −10°F, and the highest temperature is 13°F.

The difference between these two figures is calculated by subtracting. Be careful when you subtract. In particular, remember that when you see two negative signs together, you need to add. In other words, two negatives make a positive: 13 – (–10) = 13 + 10 = 23

181) The correct answer is B. You can simplify the first fraction because both the numerator and denominator are divisible by 3: $9/6 \div 3/3 = 3/2$. Then divide the denominator of the second fraction by the denominator 2 of the simplified fraction $3/2$: 10 ÷ 2 = 5. Now, multiply this number by the numerator of the first fraction to get your result: 5 × 3 = 15. You can check your answer as follows:
$9/6 = 15/10$
$9/6 \div 3/3 = 3/2$
$15/10 \div 5/5 = 3/2$

182) The correct answer is C. Remember that for questions like this one, you have to find the commonality between the facts in the question and the requested information for the solution. In this question, the commonality is the number of students. The question tells us that 17 out of every 20 students participate in a sport and that there are 800 total students. STEP 1: Determine how many groups of 20 can be formed from the total of 800: 800 ÷ 20 = 40 groups of 20 students in the school. STEP 2: To solve the problem, you then need to multiply the number of participants per group by the possible number of groups. In this problem, there are 17 participants per every group of 20. There are 40 groups of 20. So, we multiply 17 by 40 to get our answer: 17 × 40 = 680 students.

183) The correct answer is D. We have the data set: 1.6, 2.9, 4.5, 2.5, 2.5, 5.1, 5.4. The mode is the number that occurs most frequently. 2.5 occurs twice, but the other numbers only occur once. So, 2.5 is the mode.

184) The correct answer is B. We don't know the age of the 10th car, so put this in as x to solve:
(2 + 3 + 4 + 5 + 6 + 7 + 9 + 10 + 12 + x) ÷ 10 = 6
[(2 + 3 + 4 + 5 + 6 + 7 + 9 + 10 + 12 + x) ÷ 10] × 10 = 6 × 10
2 + 3 + 4 + 5 + 6 + 7 + 9 + 10 + 12 + x = 60
58 + x = 60
x = 2

185) The correct answer is C. First, multiply the erroneous average by the erroneous number of tests to get the total points: 78 × 8 = 624. Then divide this total by the correct amount: 624 ÷ 10 = 62.4

186) The correct answer is C. Education and Public Safety are the highest. So, add these two amounts together: 27% + 21% = 48%

187) The correct answer is B. The dark gray part at the bottom of each bar represents those students who will attend the dance. 45% of the freshman, 30% of the sophomores, 38% of the juniors, and 30% of the seniors will attend. Calculating the average, we get the overall percentage for all four grades: (45 + 30 + 38 + 30) ÷ 4 = 35.75%. 35% is the closest answer to 35.75%, so it best approximates our result.

188) The correct answer is B. First of all, add up the amount of faces on the chart: 4 + 3 + 2 + 3 = 12 faces. Each face represents 10 customers, so multiply to get the total number of customers: 12 × 10 =

120 customers in total for all four regions. The salespeople received $540 in total, so we need to divide this by the amount of customers: $540 ÷ 120 customers = $4.50 per customer

189) The correct answer is A. The scores were: 9.9, 9.9, 8.2, 7.6 and 6.8. Put them in ascending order and highlight the one in the middle: 6.8, 7.6, **8.2**, 9.9, 9.9

190) The correct answer is C. To find the intersection of W and X, write down both sets and highlight the numbers that they have in common: W = {2, **4**, **8**, **16**, 32}; X = {**4**, **8**, 12, **16**, 20}. So, W ∩ X = {4, 8, 16}. Then find the intersection of this new set with set Y. Set Y = {**8**, **16**, 24, 32, 40} and W ∩ X = {4, **8**, **16**}. Therefore, our answer is {8, 16}.

191) The correct answer is A. Expand by multiplying the terms as shown below:
FIRST: (**x** – 5)(**3x** + 8) = x × 3x = 3x²
OUTSIDE: (**x** – 5)(3x + **8**) = x × 8 = 8x
INSIDE: (x – **5**)(**3x** + 8) = –5 × 3x = –15x
LAST: (x – **5**)(3x + **8**) = –5 × 8 = –40
Then add all of the individual results together: 3x² + 8x + –15x + –40 = 3x² – 7x – 40

192) The correct answer is C. Isolate the integers to one side of the equation.

$$\frac{3}{4}x - 2 = 4$$

$$\frac{3}{4}x - 2 + 2 = 4 + 2$$

$$\frac{3}{4}x = 6$$

Then get rid of the fraction by multiplying both sides by the denominator.

$$\frac{3}{4}x \times 4 = 6 \times 4$$

$$3x = 24$$

Then divide to solve the problem.

$$3x \div 3 = 24 \div 3$$
$$x = 8$$

193) The correct answer is D.
Factor: $x^2 + 2x - 8 \leq 0$
$(x + 4)(x - 2) \leq 0$
Then solve each parenthetical for zero:
$(x + 4) = 0$
$-4 + 4 = 0$
$x = -4$

$(x - 2) = 0$
$2 - 2 = 0$

x = 2
So our solution is, −4 ≤ x ≤ 2

Now check. Use 0 to check to for x ≤ 2. Since 0 is actually less than 2, our proof for this should be correct.
$x^2 + 2x − 8 \leq 0$
$0 + 0 − 8 \leq 0$
$−8 \leq 0$ CORRECT

Use −5 to check for −4 ≤ x. Since −4 ≤ −5 is incorrect, our proof should also be incorrect.
$x^2 + 2x − 8 \leq 0$
$−5^2 + (2 × −5) − 8 \leq 0$
$25 − 10 − 8 \leq 0$
$25 − 18 \leq 0$
$7 \leq 0$ INCORRECT

So, we have proved that −4 ≤ x ≤ 2.

194) The correct answer is D. Notice that the equation and the inequality both contain $x − 15$. So, we can substitute y for $x − 15$ in the inequality.
$x − 15 > 0$ and $x − 15 = y$
$y > 0$

195) The correct answer is A. We know that 2 inches represents F feet. We can set this up as a ratio $2/F$. Next, we need to calculate the ratio for $F + 1$. The number of inches that represents $F + 1$ is unknown, so we will refer to this unknown as x. So we have:
$$\frac{2}{F} = \frac{x}{F+1}$$
Now cross multiply.
$$\frac{2}{F} = \frac{x}{F+1}$$
$F \times x = 2 \times (F + 1)$

$Fx = 2(F + 1)$

Then isolate x to solve.

$Fx \div F = [2(F + 1)] \div F$

$x = \frac{2(F + 1)}{F}$

196) The correct answer is D. Be careful with your zeroes. We are taking 340,000 (4 zeroes) times 1,000 (three zeroes). The result is: 340,000 × 1,000 = 340,000,000 = 34 × 10,000,000 (seven zeroes). However, our answer choices are expressed with 3.4, not 34. So, we will need to multiply by a figure with 8 zeroes to account for the change in the position of the decimal.
3.4×10^8 millimeters = 3.4 × 100,000,000 millimeters = 340,000,000

197) The correct answer is A. You should use the FOIL method in this problem. Be very careful with the negative numbers when doing the multiplication.
$2(x + 2)(x - 3) =$
$2[(x \times x) + (x \times -3) + (2 \times x) + (2 \times -3)] =$
$2(x^2 + -3x + 2x + -6) =$
$2(x^2 - 3x + 2x - 6) =$
$2(x^2 - x - 6)$

Then multiply each term by the 2 at the front of the parentheses.
$2(x^2 - x - 6) =$
$2x^2 - 2x - 12$

198) The correct answer is B. Looking at this expression, we can see that each term contains x. We can also see that each term contains y. So, first factor out xy: $2xy - 6x^2y + 4x^2y^2 = xy(2 - 6x + 4xy)$. We can also see that all of the terms inside the parentheses are divisible by 2. Now let's factor out the 2. To do this, we divide each term inside the parentheses by 2: $xy(2 - 6x + 4xy) = 2xy(1 - 3x + 2xy)$

199) The correct answer is B. First, we need to calculate the shortage in the amount of houses actually built. If H represents the amount of houses that should be built and A represents the actual number of houses built, then the shortage is calculated as: $H - A$. The company has to pay P dollars per house for the shortage, so we calculate the total penalty by multiplying the shortage by the penalty per house:
$(H - A) \times P$

200) The correct answer is B. Step 1: Apply the distributive property of multiplication by multiplying the first term in the first set of parentheses by all of the terms inside the second pair of parentheses. Then multiply the second term from the first set of parentheses by all of the terms inside the second set of parentheses.
$(5ab - 6a)(3ab^3 - 4b^2 - 3a) =$
$(5ab \times 3ab^3) + (5ab \times -4b^2) + (5ab \times -3a) + (-6a \times 3ab^3) + (-6a \times -4b^2) + (-6a \times -3a)$
Step 2: Add up the individual products in order to solve the problem:
$(5ab \times 3ab^3) + (5ab \times -4b^2) + (5ab \times -3a) + (-6a \times 3ab^3) + (-6a \times -4b^2) + (-6a \times -3a) =$
$15a^2b^4 - 20ab^3 - 15a^2b - 18a^2b^3 + 24ab^2 + 18a^2$

201) The correct answer is A. To divide, invert the second fraction and then multiply as shown.
$\frac{x}{5} \div \frac{9}{y} = \frac{x}{5} \times \frac{y}{9} = \frac{x \times y}{5 \times 9} = \frac{xy}{45}$

202) The correct answer is D. Place the integers on one side of the inequality.
$-3x + 14 < 5$
$-3x + 14 - 14 < 5 - 14$
$-3x < -9$
Then get rid of the negative number. We need to reverse the way that the inequality sign points because we are dividing by a negative.
$-3x < -9$
$-3x \div -3 > -9 \div -3$ ("Less than" becomes "greater than" because we divide by a negative number.)
$x > 3$
3.15 is greater than 3, so it is the correct answer.

203) The correct answer is A.
FIRST: $(\mathbf{x} - 2y)(\mathbf{2x^2} - y) = x \times 2x^2 = 2x^3$
OUTSIDE: $(\mathbf{x} - 2y)(2x^2 - \mathbf{y}) = x \times -y = -xy$
INSIDE: $(x - \mathbf{2y})(\mathbf{2x^2} - y) = -2y \times 2x^2 = -4x^2y$
LAST: $(x - \mathbf{2y})(2x^2 - \mathbf{y}) = -2y \times -y = 2y^2$
SOLUTION: $2x^3 + -xy + -4x^2y + 2y^2 = 2x^3 - 4x^2y + 2y^2 - xy$

204) The correct answer is A. Put in the values of 4 for x and -3 for y and simplify.
$2x^2 + 5xy - y^2 =$
$(2 \times 4^2) + (5 \times 4 \times -3) - (-3^2) =$
$(2 \times 4 \times 4) + (5 \times 4 \times -3) - (-3 \times -3) =$
$(2 \times 16) + (20 \times -3) - (9) =$
$32 + (-60) - 9 =$
$32 - 60 - 9 =$
$32 - 69 = -37$

205) The correct answer is C.
$6 + 8(2\sqrt{x} + 4) = 62$
$6 - 6 + 8(2\sqrt{x} + 4) = 62 - 6$
$8(2\sqrt{x} + 4) = 56$
$16\sqrt{x} + 32 = 56$
$16\sqrt{x} + 32 - 32 = 56 - 32$
$16\sqrt{x} = 24$
$16\sqrt{x} \div 16 = 24 \div 16$
$\sqrt{x} = 24 \div 16$
$\sqrt{x} = \dfrac{24}{16}$
$\sqrt{x} = \dfrac{24 \div 8}{16 \div 8} = \dfrac{3}{2}$

206) The correct answer is D. $\sqrt{18} \times \sqrt{8} = \sqrt{18 \times 8} = \sqrt{144} = \sqrt{12 \times 12} = 12$

207) The correct answer is A. Perform the multiplication on the terms in the parentheses.
$2(3x - 1) = 4(x + 1) - 3$
$6x - 2 = (4x + 4) - 3$
Then simplify.
$6x - 2 = (4x + 4) - 3$
$6x - 2 = 4x + 1$
$6x - 2 - 1 = 4x + 1 - 1$
$6x - 3 = 4x$
Then isolate x to get your answer.
$6x - 3 = 4x$
$6x - 4x - 3 = 4x - 4x$
$2x - 3 = 0$
$2x - 3 + 3 = 0 + 3$

$2x = 3$
$2x \div 2 = 3 \div 2$
$x = {}^3/_2$

208) The correct answer is C. The first point on the graph lies at $x = 10$, so we can eliminate answer choices A and B. The point for the y coordinate that corresponds to $x = 10$ is 63 not 68, so we can eliminate answer choice D.

209) The correct answer is B. The area of a triangle is base times height divided by 2. First, calculate the area of triangle FGJ: $[6 \times (8 + 10)] \div 2 = (6 \times 18) \div 2 = 108 \div 2 = 54$. Then, calculate the area of triangle FGH: $(6 \times 8) \div 2 = 24$. The area of triangle FHJ is calculated by subtracting the area of triangle FGH from the area of triangle FGJ: $54 - 24 = 30$

210) The correct answer is B. The measurement of a straight line is 180° so the measurement of angle A is $180° - 109° = 71°$. Since this is an isosceles triangle, angle A and angle B are equal. The sum of the degrees of the three angles of any triangle is 180°, so we subtract to find the measurement of angle A: $180° - 71° - 71° = 38°$.

211) The correct answer is D. Area of a circle = $\pi \times radius^2$, and radius = diameter ÷ 2. Our diameter is 36, so the radius is 18. Therefore, the area is: $\pi \times radius^2 = \pi \times 18^2 = 324\pi$

212) The correct answer is D. Isolate the whole numbers to one side of the equation first.

$20 - \frac{3x}{4} \geq 17$

$(20 - 20) - \frac{3x}{4} \geq 17 - 20$

$-\frac{3x}{4} \geq -3$

Then get rid of the fraction.

$-\frac{3x}{4} \geq -3$

$\left(4 \times -\frac{3x}{4}\right) \geq -3 \times 4$

$-3x \geq -12$

Then deal with the remaining whole numbers.
$-3x \geq -12$
$-3x \div -3 \geq -12 \div -3$
$x \leq 4$
Remember to reverse the way the sign points when you divide by a negative number.

213) The correct answer is D. The volume of a box is calculated by taking the length times the width times the height: $5 \times 6 \times 10 = 300$

214) The correct answer is B. Substitute the values into the equation to solve. For $x = 2$ and $y = 3$:
$10x + 3y = (10 \times 2) + (3 \times 3) = 20 + 9 = 29$

215) The correct answer is B. The formula for perimeter is as follows: P = 2W + 2L. The patch is 12 yards by 10 yards, so we need 12 yards × 2 for the long sides patch and 10 yards × 2 for the shorter sides of the patch: (2 × 10) + (2 × 12) = 20 + 24 = 44

216) The correct answer is B. Cone volume = (π × radius² × height) ÷ 3
volume = (π3² × 4) ÷ 3 = (π9 × 4) ÷ 3 = π36 ÷ 3 = 12π

217) The correct answer is C. Use the Pythagorean theorem for the hypotenuse length $C = \sqrt{A^2 + B^2}$
$\sqrt{7^2 + B^2} = 14$
$\left(\sqrt{7^2 + B^2}\right)^2 = 14^2$
$7^2 + B^2 = 196$
$49 + B^2 = 196$
$B^2 = 196 - 49$
$B^2 = 147$
$B = \sqrt{147}$

218) The correct answer is A. We start off with point B, which is represented by the coordinates (0, 2). The line is then shifted 5 units to the left and 4 units up. When we go to the left, we need to deduct the units, and when we go up we need to add units. So, do the operations on each of the coordinates in order to solve: 0 – 5 = –5 and 2 + 4 = 6, so our new coordinates are (–5, 6).

219) The correct answer is D. The string that goes around the front, back, and sides of the package is calculated as follows: 20 + 10 + 20 + 10 = 60. The string that goes around the top, bottom, and sides of the package is calculated in the same way since the top and bottom are equal in length to the front and back: 20 + 10 + 20 + 10 = 60. So, 120 inches of string is needed so far. Then, we need 15 extra inches for the bow: 120 + 15 = 135

220) The correct answer is C. An equilateral triangle has three equal sides and three equal angles. Since all 3 angles in any triangle need to add up to 180 degrees, each angle of an equilateral triangle is 60 degrees (180 ÷ 3 = 60). Angles that lie along the same side of a straight line must add up to 180. So, we calculate angle *a* as follows: 180 – 60 = 120

221) The correct answers is D. The area of a circle is: πR^2. The area of circle A is $\pi \times 5^2 = 25\pi$ and the area of circle B is $\pi \times 3^2 = 9\pi$. So, the difference between the areas is 16π. The formula for circumference is: $\pi 2R$. The circumference of circle A is $\pi \times 2 \times 5 = 10\pi$ and the circumference for circle B is $\pi \times 2 \times 3 = 6\pi$. The difference in the circumferences is 4π. So, answer D is correct.

222) The correct answer is A. You need to use the distance formula: $d = \sqrt{(x_2 - x_1)^2 + (y_2 - y_1)^2}$
Put in the values provided, which were $(4\sqrt{7}, -2)$ and $(7\sqrt{7}, 4)$. Then multiply and simplify to solve.
$\sqrt{(x_2 - x_1)^2 + (y_2 - y_1)^2} =$
$\sqrt{(7\sqrt{7} - 4\sqrt{7})^2 + (4 - -2)^2} =$
$\sqrt{(3\sqrt{7})^2 + (6)^2} =$

$\sqrt{(9 \times 7) + 36} =$

$\sqrt{63 + 36} =$

$\sqrt{99} = \sqrt{9 \times 11} = 3\sqrt{11}$

223) The correct answer is A. y is negative when x is negative, and y is positive when x is positive. Looking at the value for x = 10, we can see that the output for y is 0.10 when x = 10, so the correct function is $f(x) = \frac{1}{x}$

224) The correct answer is D. First, solve for the function in the inner-most set of parentheses, in this case $f_1(x)$. To solve, you simply have to look at the first table. Find the value of 2 in the first column and the related value in the second column. For x = 2, $f_1(2)$ = 5. Then, take this new value to solve for $f_2(x)$. Look at the second table. Find the value of 5 in the first column and the related value in the second column. For x = 5, $f_2(5)$ = 25.

225) The correct answer is C. Two whole numbers that are greater than 1 will always result in a greater number when they are multiplied by each other, rather than when those numbers are divided by each other or subtracted from each other. So, for positive integers, x × y will always be greater than the following:

x ÷ y
y ÷ x
x – y
y – x
1 ÷ x
1 ÷ y

226) The correct answer is D. Sine is "opposite over hypotenuse," so the sine of angle Z is XY divided by XZ.
Substitute values into the formula:
sin z = $^{XY}/_{XZ}$
sin z = $^{XY}/_{10}$
0.643 = $^{XY}/_{10}$
0.643 × 10 = $^{XY}/_{10}$ × 10
0.643 × 10 = XY
6.43 = XY

227) The correct answer is B. The trig formulas state that cos A° = sin(90° – A°). This is a 41 degree angle, so sin 41° = cos(90° – 41°) = cos 49°. The sine of a 41 degree angle is 0.656059, and the cosine of a 49 degree angle is also 0.656059.

228) The correct answer is B. We use tangent to solve this problem because we are dealing with the opposite and adjacent sides.
tan 70° = BA ÷ PB ("opposite over adjacent")

Now substitute the value for line segment BA.
tan 70° = 72 ÷ PB

Then simplify.
tan 70° × PB = (72 ÷ PB) × PB
tan 70° × PB = 72
(tan 70° × PB) ÷ tan 70° = 72 ÷ tan 70°
PB = 72 ÷ tan 70°

229) The correct answer is D. The trig formulas state that $\sin^2 A = 1 - \cos^2 A$. The problem tells us that the cosine squared of angle A is 0.235040368, so put the value into the formula to solve.
$\sin^2 A = 1 - \cos^2 A$
$\sin^2 A = 1 - 0.235040368$
$\sin^2 A = 0.764959632$

230) The correct answer is D. We want to calculate the length of side BC, which is the opposite side of this triangle. We have the length of the hypotenuse, so we can determine the length of side BC as follows:
sin A = BC/AC ("opposite over hypotenuse")
sin A = BC/14
0.8192 = BC/14
0.8192 × 14 = BC
11.469 = BC

231) The correct answer is B. The number of deserts is D and the number of main dishes is M. There are 4 family members, so both D and M are 4.
(D × ?) + (M × $8) = $48
(4 × ?) + (4 × $8) = $48
(4 × ?) + $32 = $48
(4 × ?) + $32 − $32 = $48 − $32
4 × ? = $16
(4 × ?) ÷ 4 = $16 ÷ 4
? = $4

232) The correct answer is C. You need to evaluate the equation in order to determine which operations you need to perform on any new equation containing the operation ⊖ and variables x and y. For the special operation $(x \ominus y) = (5x + 2y)$, in any new equation: Operation ⊖ is addition; the number or variable before ⊖ is multiplied by 5; the number or variable after ⊖ is multiplied by 2.
So, the new equation (6 ⊖ z) = 8 becomes (6 × 5) + (z × 2) = 44
Now solve.
(6 × 5) + (z × 2) = 44
30 + (z × 2) = 44
30 − 30 + (z × 2) = 44 − 30
z × 2 = 14
z = 7

233) The correct answer is B. Substitute $x + 3$ for x in the original function to solve. So, $x^2 + 3x - 8$ becomes $(x+3)^2 + 3(x+3) - 8$

234) The correct answer is C. Negative numbers do not have square roots that are real numbers. So, we need to use an imaginary number to solve this problem.

235) The correct answer is B. Divide the capacity by the time in order to get the rate: 1200 gallons ÷ 75 minutes = 16 gallons per minute.

236) The correct answer A. The range is all of the possible y values for the function. The output of the function will be zero when $x = -2$ since $-5(-2)^3 - 40 = (-5 \times -8) - 40 = 40 - 40 = 0$. The output of the function will be positive when $x < -2$ and negative when $x > -2$. So, the range is all real numbers.

237) The correct answer is C. Since there are 60 seconds in a minute, and heartbeats are measured in 10 second units, we divide the seconds as follows: 60 ÷ 10 = 6. Accordingly, the BPM is calculated by talking B times 6: BPM = B6.

238) The correct answer is D. In order to find the excess amount, we deduct the ideal BPM of 60 from the patient's actual BPM: BPM − 60

239) The correct answer is B. The two projects are being given different weights, so each project needs to have its own variable. Project X counts for 45% of the final grade, so the weighted value of project X is .45X. Project Y counts for 55%, so the weighted value of project Y is .55Y. The final grade is the total of the values for the two projects. So, we add to get our equation: .45X + .55Y

240) The correct answer is D. Set up each part of the problem as an equation. The museum had twice as many visitors on Tuesday (T) as on Monday (M), so T = 2M. The number of visitors on Wednesday exceeded that of Tuesday by 20%, so W = 1.20 × T. Then express T in terms of M for Wednesday's visitors: W = 1.20 × T = 1.20 × 2M = 2.40M. Finally, add the amounts together for all three days: M + 2M + 2.40M = 5.4M

Solutions and Explanations for Practice Test Set 4 – Questions 241 to 320

241) The correct answer is C. Yesterday the train traveled $117^{3}/_{4}$ miles, and today it traveled $102^{1}/_{6}$ miles. To find the difference, we subtract these two amounts. Because the fraction on the first mixed number is greater than the fraction on the second mixed number, we can subtract the whole numbers and the fractions separately: $117^{3}/_{4}$ miles − $102^{1}/_{6}$ miles = ? STEP 1: Subtract the whole numbers: 117 − 102 = 15 miles. STEP 2: Perform the operation on the fractions by finding the lowest common denominator. $^{3}/_{4}$ miles − $^{1}/_{6}$ miles = ? In order to find the LCD, we would normally need to find the common factors first. Our denominators in this problem are 4 and 6. The factors of 4 are: 1 × 4 = 4; 2 × 2 = 4. The factors of 6 are: 1 × 6 = 6; 2 × 3 = 6. We do not have two factors in common, so we know that we need to find a new denominator which is greater than 6. In this problem, the LCD is 12 since 3 × 4 = 12 and 2 × 6 = 12. So, we express the fractions $^{3}/_{4}$ miles + $^{1}/_{6}$ miles in their LCD form: $^{3}/_{4} \times ^{3}/_{3} = ^{9}/_{12}$ and $^{1}/_{6} \times ^{2}/_{2} = ^{2}/_{12}$. Then subtract these two fractions: $^{9}/_{12} - ^{2}/_{12} = ^{7}/_{12}$. STEP 3: Combine the results from the two previous steps to solve the problem: $117^{3}/_{4}$ miles − $102^{1}/_{6}$ miles = $15^{7}/_{12}$ miles.

242) The correct answer is D. Sam is driving at 70 miles per hour, and at 10:30 am he is 140 miles from Farnam. STEP 1: We need to find out how far he will be from Farnam at 11:00 am, so we need to work

out how far he will travel in 30 minutes. STEP 2: If Sam is traveling at 70 miles an hour, then he travels 35 minutes in half an hour: 70 miles in one hour × $\frac{1}{2}$ hour = 35 miles. STEP 3: If he was 140 miles from Farnam at 10:30 am, he will be 105 miles from Farnam at 11:00 am: 140 − 35 = 105 miles

243) The correct answer is C. The twelve students who failed the test represent one-third of the class. Since one-third of the students have failed, we can think of the class as being divided into three groups: Group 1: The 12 students who failed; Group 2: 12 students who would have passed; Group 3: 12 more students who would have passed. So, the class consists of 36 students in total. In other words, we need to multiply by three to find the total number of students: 12 × 3 = 36

244) The correct answer is B. The problem tells us that sales this week were $600 and sales last week were $525. STEP 1: First, we need to find the difference in sales between the two weeks: $600 - $525 = $75 more in sales this week. STEP 2: Since each book is sold for $5, we divide this figure into the total in order to find out how many books were sold: $75 more sales ÷ $5 per book = 15 more books sold this week.

245) The correct answer is D. A percentage can always be expressed as a number with two decimal places. For example, 15% = 0.15 and 20% = 0.20. In our problem, 16% = 0.16. So D is correct.

246) The correct answer is B. STEP 1: First of all, you need to calculate the amount of the discount: $18 original price × 40% = $18 × 0.40 = $7.20 discount. STEP 2: Then deduct the amount of the discount from the original price to calculate the sales price of the item: $18 original price - $7.20 discount = $10.80 sales price.

247) The correct answer is B. If David answered 18 questions incorrectly on the exam and lost 36 points, and he then earned 25 extra credit points, his score was lowered by 11 points. STEP 1: To do the calculation, we need to take the points lost on the exam and add the extra credit points: −36 + 25 = −11. STEP 2: Since the question is asking how much the score was lowered, you need to give the amount as a positive number.

248) The correct answer is C. STEP 1: Calculate the amount of the tax increase: $480 × 7.5% = ? $480 original tax amount × 0.075 = $36 proposed increase in tax. STEP 2: Then add the increase to the original amount to get the amount of the tax after the proposed increase: $480 original tax + $36 increase in tax = $516 tax after increase

249) The correct answer is B. If there are 12 children and each one is supposed to receive 4 items, we can do the calculation as follows: 12 children × 4 items per child = 48 items required in total. Now subtract the total from the amount she already has in order to determine how many more she needs: 48 items required in total − 40 items available = 8 items still needed.

250) The correct answer is C. First of all, you have to find out how many students were absent on Tuesday. To find the number of absent students, you have to multiply the total number of students in the class by the percentage of the absence for Tuesday: 20 students in total × 5% = 1 student absent on Tuesday. Now calculate the absences for Wednesday in the same way: 20 students in total × 20% = 4 students absent on Wednesday. The problem is asking you how many more students were absent on Wednesday than Tuesday, so you need to subtract the two figures that you have just calculated.

4 students absent on Wednesday – 1 student absent on Tuesday = 3 students. So, 3 more students were absent on Wednesday.

251) The correct answer is B. The problem is asking you for the amount that the number of births per hospital in Johnson County for 2016 exceeded those for 1998. STEP 1: First we have to calculate the amount for 2016. In order to calculate this figure, you have to divide the total births by the number of hospitals in each data set. For 2016, we have 240 total births and 15 hospitals in the data set: 240 ÷ 15 = 16 births per hospital for 2016. STEP 2: Now calculate the amount for 1998. In our problem, this amount is provided. We can see that there were 12 births per hospital in Johnson County in 1998. STEP 3: Now subtract the amounts for the two years to get your answer: 16 – 12 = 4 more births per hospital in 2016.

252) The correct answer is D. STEP 1: You need to multiply the number of miles that she is going to travel by the amount of time it takes her to travel one mile: 17 minutes for 1 mile × 5 miles to travel = 85 minutes needed. STEP 2: Now express the result in hours and minutes, remembering of course that an hour has 60 minutes: 85 minutes – 60 minutes = 25 minutes left. So, the answer is 1 hour and 25 minutes.

253) The correct answer is D. Sam's final grade for a class is based on his scores from a midterm test (M), a project (P), and a final exam (F), but the midterm test counts twice as much as the project, and the final exam counts twice as much as the midterm. Therefore, we have to count variable M twice. The value of the midterm is doubled and variable F is double of the midterm, so we have to count variable F 4 times. So, the equation is: P + 2M + 4F

254) The correct answer is D. The problem tells us that Bart rides at a rate of 12 miles per hour. We also know that he arrives in the town of Wilmington at 3:00 pm. The question is asking us how far Bart will be from Mount Pleasant at 5:00 pm. STEP 1: Calculate the time difference: 5:00 pm – 3:00 pm = 2 hours difference. STEP 2: Calculate the distance traveled: 12 miles per hour × 2 hours = 24 miles traveled. STEP 3: Calculate the distance left. The town of Mount Pleasant is 50 miles from Wilmington: 50 miles to travel – 24 miles traveled = 26 miles left.

255) The correct answer is C. The ticket office sold 360 more tickets on Friday than it did on Saturday. The office sold 2570 tickets in total during Friday and Saturday. STEP 1: Subtract the excess: 2570 – 360 = 2210. STEP 2: Allocate the above figure to each day: 2210 ÷ 2 = 1105. STEP 3: Calculate Friday's amount by adding back in the excess: 1105 + 360 = 1465.

256) The correct answer is C. STEP 1: Calculate the beginning height in inches. Remember that there are 12 inches in a foot: 5 feet × 12 inches per foot = 60 inches in height. STEP 2: Calculate the increase in height: 60 inches × 10% = 6 inches. STEP 3: Calculate the new height by adding the increase to the number at the beginning: 5 feet + 6 inches = 5 feet 6 inches.

257) The correct answer is D. STEP 1: Calculate the amount of money spent on the original purchase of the jeans: 2 × $22.98 = $45.96. STEP 2: Calculate the value of the items acquired in the exchange, which in this case, is the value of the sweaters: 3 × $15.50 = $46.50. STEP 3: Calculate the difference between the value of the items acquired and the amount of money originally spent. Value of the items acquired: 3 × $15.50 = $46.50. Amount of money originally spent: 2 × $22.98 = $45.96. Difference: (3 × $15.50) – (2 × $22.98).

258) The correct answer is D. We know that we have to round to the nearest hundredth. The hundredth decimal place is the number 2 positions to the right of the decimal. For example, .01 is 1 one hundredth. In our question, the first jump of 3.246 is rounded up to 3.25. The second jump of 3.331 is rounded down to 3.33. The third jump of 3.328 is rounded up to 3.33. Then add these three figures together to get your answer: 3.25 + 3.33 + 3.33 = 9.91

259) The correct answer is D. You have to find the relationship between the number given in each row in the left column and the corresponding number in the right column. "9:50 am to 10:36 am" represents a journey time of 46 minutes. 11:15 to 12:01 is also 46 minutes, and so on. If we go 46 minutes back from 5:51 pm, we get 5:05 pm for our answer.

260) The correct answer is B. He owns 26 yachts and needs 6 feet 10 inches of rope for each one. Convert the feet and inches measurement to inches: 6 feet 10 inches = (6 × 12) + 10 inches = 72 + 10 = 82 inches. Then multiply buy the number of items: 26 × 82 = 2132 inches of rope needed. Then convert back to feet and inches: 2132 inches ÷ 12 = 177 feet 8 inches.

261) The correct answer is C. STEP 1: You can simplify the first fraction because both the numerator and denominator are divisible by 3: $^3/_6 \div ^3/_3 = ^1/_2$. STEP 2: Then divide the denominator of the second fraction ($^x/_{14}$) by the denominator of the simplified fraction ($^1/_2$) from above: 14 ÷ 2 = 7. STEP 3: Now, multiply the number from step 2 by the numerator of the fraction we calculated in step 1 in order to get your result: 1 × 7 = 7. You can check your answer as follows: $^3/_6 = ^7/_{14}$; $^3/_6 \div ^3/_3 = ^1/_2$; $^7/_{14} \div ^7/_7 = ^1/_2$

262) The correct answer is D. This problem is asking for the ratio of non-faulty mp3 players to the quantity of faulty mp3 players. Therefore, you must put the quantity of non-faulty mp3 players before the colon in the ratio. In this problem, 1% of the players are faulty. 1% × 100 = 1 faulty player in every 100 players. 100 − 1 = 99 non-faulty players. So, the ratio is 99:1. As explained previously, the number before the colon and the number after the colon can be added together to get the total quantity.

263) The correct answer is A. Step 1 – Determine the arithmetic mean for the prices: 12 + 14 + 10 + 8 = 44 ÷ 4 = 11. Step 2 – Calculate the "difference from the mean" for each price. Price 1: 12 − 11 = 1; Price 2: 14 − 11 = 3; Price 3: 10 − 11 = −1; Price 4: 8 − 11 = −3. Step 3 – Square the "difference from the mean" for each score. Price 1: $1^2 = 1$; Price 2: $3^2 = 9$; Price 3: $−1^2 = 1$; Price 4: $−3^2 = 9$. Step 4 – Find the mean of the squared figures to get the variance. 1 + 9 + 1 + 9 = 20 ÷ 4 = 5

264) The correct answer is A. Find the total of the items in the sample space: 5 + 10 + 8 + 12 = 35. We want to know the chance of getting an orange balloon, so put that in the denominator: $\frac{10}{35} = \frac{2}{7}$

265) The correct answer is D. We have 54 cards in the deck (13 × 4 = 52). We have taken out two spades, one heart, and a club, thereby removing 4 cards. So, the available data set is 48 (52 − 4 = 48). The desired outcome is drawing a heart. We have 13 hearts to begin with and one has been removed, so there are 12 hearts left. So, the probability of drawing a heart is $^{12}/_{48} = ^1/_4$

266) The correct answer is D. Look at the bars for June 1 at the far right side of the graph. First, find the total amount of accidents on that date. Cars were involved in 30 accidents, vans in 20 accidents, pick-ups in 10 accidents, and SUV's in 5 accidents. So, there were 65 accidents in total (30 + 20 + 10 + 5 = 65). Then divide the number of accidents for pick-ups and vans into the total: 30 ÷ 65 = 46.1538% ≈ 46%

267) The correct answer is C. There are 2 stars for speeding, and each star equals 30 violations, so there were 60 speeding violations in total. The fine for speeding is $50 per violation, so the total amount collected for speeding violations was: 60 speeding violations × $50 per violation = $3000. There are three stars for other violations, which is equal to 90 violations (3 × 30 = 90). Other violations are $20 each, so the total for other violations is: 90 × $20 = $1800. Next, we need to deduct these two amounts from the total collections of $6,000 to find the amount collected for parking violations: $6000 − $3000 − $1800 = $1200 in total for parking violations. There is one star for parking violations, so there were 30 parking violations. We divide to get the answer: $1200 income for parking violations ÷ 30 parking violations = $40 each

268) The correct answer is D. The most striking relationship on the graph is the line for ages 65 and over, which clearly shows a negative relationship between exercising outdoors and the number of days of rain per month. You will recall that a negative relationship exists when an increase in one variable causes a decrease in the other variable. So, we can conclude that people aged 65 and over seem less inclined to exercise outdoors when there is more rain.

269) The correct answer is A. The range is the highest amount minus the lowest amount: 21 − 3 = 18

270) The correct answer is B. Two members have lost 12 kilograms, and all of the other amounts occur only one time each. So, 12 is the mode.

271) The correct answer is A.
Factor: $x^2 + 4x + 3 > 0$
$(x + 1)(x + 3) > 0$
Then solve each parenthetical for zero:
$(x + 1) = 0$
$−1 + 1 = 0$
$x = −1$

$(x + 3) = 0$
$−3 + 3 = 0$
$x = −3$
So, $x < −3$ or $x > −1$

Now check. Use 0 to check to for $x > −1$. Since $0 > −1$ is correct, our proof for this should also be correct.
$x^2 + 4x + 3 > 0$
$0 + 0 + 3 > 0$
$3 > 0$ CORRECT

Use −2 to check for $x < −3$. Since $−2 < −3$ is incorrect, our proof should also be incorrect.
$x^2 + 4x + 3 > 0$
$−2^2 + (4 × −2) + 3 > 0$
$4 − 8 + 3 > 0$
$−4 + 3 > 0$
$−1 > 0$ INCORRECT

Therefore, we have checked that $x < −3$ or $x > −1$

272) The correct answer is C.
FIRST: $(\mathbf{x} - 9y)(\mathbf{x} - 9y) = x \times x = x^2$
OUTSIDE: $(\mathbf{x} - 9y)(x - \mathbf{9y}) = x \times -9y = -9xy$
INSIDE: $(x - \mathbf{9y})(\mathbf{x} - 9y) = -9y \times x = -9xy$
LAST: $(x - \mathbf{9y})(x - \mathbf{9y}) = -9y \times -9y = 81y^2$
SOLUTION: $x^2 - 18xy + 81y^2$

273) The correct answer is B. Deal with the whole numbers first.
$6 + \frac{x}{4} \geq 22$
$6 - 6 + \frac{x}{4} \geq 22 - 6$
$\frac{x}{4} \geq 16$

Then eliminate the fraction.
$\frac{x}{4} \geq 16$
$4 \times \frac{x}{4} \geq 16 \times 4$
$x \geq 64$

274) The correct answer is A. Perform long division of the polynomial.
```
             x + 3
       _____
x – 4) x² – x – 12
       x² – 4x
       _____
            3x – 12
            3x – 12
            _____
                 0
```

275) The correct answer is A. Factor out xy: $18xy - 24x^2y - 48y^2x^2 = xy(18 - 24x - 48xy)$
Then, factor out the common factor of 6: $xy(18 - 24x - 48xy) = 6xy(3 - 4x - 8xy)$

276) The correct answer is C. Multiply the integers and add the exponents on the variables:
$\sqrt{15x^3} \times \sqrt{8x^2} =$
$\sqrt{15x^3 \times 8x^2} =$
$\sqrt{15 \times 8 \times x^3 \times x^2} =$
$\sqrt{120x^5} = \sqrt{2 \times 2 \times x^2 \times x^2 \times x \times 30} = 2x^2\sqrt{30x}$

277) The correct answer is B. We know from the second equation that y is equal to $x + 7$. So put $x + 7$ into the first equation for the value of y to solve.
$-3x - 1 = y$
$-3x - 1 = x + 7$
$-3x - 1 + 1 = x + 7 + 1$
$-3x - x = x - x + 8$
$-4x = 8$
$-4x \div -4 = 8 \div -4$
$x = -2$

Now we know that the value of x is –2, so we can put that into the equation to solve for y.
–3x – 1 = y
(–3 × –2) – 1 = y
6 – 1 = y
y = 5

278) The correct answer is D. Any negative exponent is equal to 1 divided by the variable. So, $x^{-4} = 1 \div x^4$

279) The correct answer is C. Deal with the integers that are outside the parentheses first. Then remove the radical to solve.
$5(4\sqrt{x} - 8) = 40$
$20\sqrt{x} - 40 = 40$
$20\sqrt{x} - 40 + 40 = 40 + 40$
$20\sqrt{x} = 80$
$20\sqrt{x} \div 20 = 80 \div 20$
$\sqrt{x} = 4$
$\sqrt{x}^2 = 4^2$
$x = 16$

280) The correct answer is A. Find the cube roots of the integers and factor them. Express the result as a rational number.
$$\sqrt[3]{\frac{8}{27}} = \sqrt[3]{\frac{2 \times 2 \times 2}{3 \times 3 \times 3}} = \frac{2}{3}$$

281) The correct answer is D. When you have fractions in the numerator and denominator of another fraction, you can divide the two fractions as follows:
$$\frac{5a/b}{2a/a-b} = \frac{5a}{b} \div \frac{2a}{a-b}$$

Then invert and multiply just like you would for any other fraction.
$$\frac{5a}{b} \div \frac{2a}{a-b} = \frac{5a}{b} \times \frac{a-b}{2a} = \frac{5a^2 - 5ab}{2ab}$$

Then simplify, if possible.
$$\frac{5a^2 - 5ab}{2ab} = \frac{a(5a - 5b)}{a(2b)} = \frac{\cancel{a}(5a - 5b)}{\cancel{a}(2b)} = \frac{5a - 5b}{2b}$$

282) The correct answer is B. The line that represents the diameter of the circle forms the hypotenuse of a triangle. Side A of the triangle begins on (0, 0) and ends on (0, 2), with a length of 2. Side B of the triangle begins on (0, 2) and ends on (2, 2), so it also has a length of 2. So, the diameter of the circle is: $\sqrt{2^2 + 2^2} = \sqrt{8} = 2\sqrt{2}$. Next, we need to calculate the radius of the circle. The radius of the circle is $\sqrt{2}$ because the diameter is $2\sqrt{2}$ and the formula for the radius of a circle is ½ × diameter = radius. Finally, we can use the formula for the area of a circle to solve the problem: $\pi\sqrt{2}^2 = 2\pi$

283) The correct answer is D. A negative linear relationship exists when an increase in one variable results in a decrease in the other variable. This is represented by chart D.

284) The correct answer is D. We can determine the function that this graph represents to solve the problem. First, we can perform division to determine that the plane travels 6.5 miles per minute. For example, the line for 120 minutes is at 780 miles: 780 miles ÷ 120 minutes = 6.5 miles per minute. Since the plane is travelling at a constant rate, the graph above expresses the function: $f(x) = x \times 6.5$. The domain represents the x values on the graph and the range represents the y values, so answer D is correct,

285) The correct answer is B. Triangle area = (base × height) ÷ 2 = (4 × 15) ÷ 2 = 60 ÷ 2 = 30

286) The correct answer is C. Use the formula for circumference: (2 × π × radius). So, we calculate the circumference of the large circle as: 2 × π × 8 = 16π. The circumference of the small circle is: 2 × π × 5 = 10π. Then, we subtract to get our solution: 16π − 10π = 6π

287) The correct answer is A. An isosceles triangle has two equal sides, so answer A is correct. If an altitude is drawn in an isosceles triangle, we have to put a straight line down the middle of the triangle from the peak to the base. Dividing the triangle in this way would form two right triangles, rather than two equilateral triangles. So, answer B is incorrect. The base of an isosceles triangle can be longer than the length of each of the other two sides, so answer C is incorrect. The sum of all three angles of any triangle must be 180 degrees, rather than 360 degrees. So, answer D is incorrect.

288) The correct answer is A. We need to calculate the radius of the shaded portion. Since the height of the shaded portion is 6 and the height of the entire cone is 18, we know by using the rules of similarity that the ratio of the radius of the shaded portion to the radius of the entire cone is $6/18$ or $1/3$. Using this fraction, we can calculate the radius for the shaded portion. The radius of the entire cone is 9, so the radius of the shaded portion is 3: 9 × $1/3$ = 3. Then, calculate the volume of the shaded portion:
$(\pi \times 3^2 \times 6) \div 3 = 54\pi \div 3 = 18\pi$

289) The correct answer is D. One side of the triangle is 18 meters and the other side of the triangle is 30 meters, so we can put these values into the Pythagorean Theorem in order to solve the problem.
$\sqrt{A^2 + B^2} = C$
$\sqrt{18^2 + 30^2} = C$
$\sqrt{324 + 900} = C$
$\sqrt{1224} = C$
35 × 35 = 1225
So, the square root of 1224 is approximately 35.

290) The correct answer is B. We can see that when x = 80, y = 60. So, when x = 160, y = 120. Alternatively, if you prefer, you can determine that the line represents the function: $f(x) = x \times 0.75$. Then substitute 160 for x: $x \times 0.75 = 160 \times 0.75 = 120$

291) The correct answer is C. Perform the operation inside the absolute value signs then make the result negative since there is a negative sign in front of the absolute value: − | 5 − 8| = − | −3| = − |3| = −3

292) The correct answer is A. First, calculate the total square footage available. There are 4 areas that are 10 by 10 each, so we have this equation: 4 × (10 × 10) = 400 square feet in total. Then calculate the square footage of the new offices: 20 × 10 = 200 and 2 offices × (10 × 8) = 160; 200 + 160 = 360 total square feet for the new offices. So, the remaining square footage for the common area is determined by taking the total square footage minus the square footage of the new offices: 400 − 360 = 40 square feet remaining. Since each existing office is 10 feet long, we know that the new common area needs to be 10 feet long in order to fit in. So, the new common area is 4 feet × 10 feet.

293) The correct answer is C. Use the Pythagorean Theorem to solve. $S = \sqrt{Q^2 + R^2}$
$S = \sqrt{Q^2 + R^2} = \sqrt{3^2 + 2^2} = \sqrt{9 + 4} = \sqrt{13}$

294) The correct answer is C. Calculate the area for each of the cupboards: 8 × 2 = 16 and 5 × 2 = 10. Therefore, the total area for both cupboards is 16 + 10 = 26. Then find the area for the entire kitchen: 8 × 12 = 96. Then deduct the cupboards from the total: 96 − 26 = 70

295) The correct answer is A. First, we need to find the circumference of the semicircle on the left side of the figure. The width of the rectangle of 10 inches forms the diameter of the semicircle, so the circumference of an entire circle with a diameter of 10 inches would be 10π inches. We need the circumference for a semicircle only, which is half of a circle, so we need to divide the circumference by 2: 10π ÷ 2 = 5π. Since the right side of the figure is an equilateral triangle, the two sides of the triangle have the same length as the width of the rectangle, so they are 10 inches each. Finally, you need to add up the lengths of all of the sides to get the answer: 18 + 18 + 10 + 10 + 5π = 56 + 5π inches

296) The correct answer is D. To solve the problem, insert the values provided into the formula for the volume of a pyramid: $\frac{1}{3}$ × length × width × height

$\frac{1}{3}$ × length × width × height = 30

$\frac{1}{3}$ × 5 × 3 × height = 30

$\frac{15}{3}$ × height = 30

5 × height = 30

5 ÷ 5 × height = 30 ÷ 5

height = 6

297) The correct answer is C.
Here is the solution for *y* intercept:
$5x^2 + 4y^2 = 120$
$5(0)^2 + 4y^2 = 120$
$0 + 4y^2 = 120$
$4y^2 = 120$
$4y^2 ÷ 4 = 120 ÷ 4$
$y^2 = 30$

$y = \sqrt{30}$
So, the y intercept is $(0, \sqrt{30})$

Here is the solution for x intercept:
$5x^2 + 4y^2 = 120$
$5x^2 + 4(0)^2 = 120$
$5x^2 + 0 = 120$
$5x^2 = 120$
$5x^2 \div 5 = 120 \div 5$
$x^2 = \sqrt{24}$
So the x intercept is $(\sqrt{24}, 0)$

298) The correct answer is B. Use the slope-intercept formula to calculate the slope: $y = mx + b$, where m is the slope and b is the y intercept. In our question, $x = 4$ and $y = 15$. The line crosses the y axis at 3, so put the values into the formula.
$y = mx + b$
$15 = m4 + 3$
$15 - 3 = m4 + 3 - 3$
$12 = m4$
$12 \div 4 = m$
$3 = m$

299) The correct answer is A. The perimeter of a rectangle is equal to two times the length plus two times the width. We can express this concept as an equation: $P = 2L + 2W$. Now set up formulas for the perimeters both before and after the increase.
STEP 1 – Before the increase:
$P = 2L + 2W$
$48 = 2L + 2W$
$48 \div 2 = (2L + 2W) \div 2$
$24 = L + W$
$24 - W = L + W - W$
$24 - W = L$
STEP 2 – After the increase (length is increased by 5 and width is doubled):
$P = 2L + 2W$
$92 = 2(L + 5) + (2 \times 2)W$
$92 = 2L + 10 + 4W$
$92 - 10 = 2L + 10 - 10 + 4W$
$82 = 2L + 4W$
Then solve by substitution. In this case, we substitute $24 - W$ (which we calculated in the "before" equation in step 1) for L in the "after" equation calculated in step 2, in order to solve for W.
$82 = 2L + 4W$
$82 = 2(24 - W) + 4W$
$82 = 48 - 2W + 4W$
$82 - 48 = 48 - 48 - 2W + 4W$
$82 - 48 = -2W + 4W$
$34 = -2W + 4W$

34 = 2W
34 ÷ 2 = 2W ÷ 2
17 = W
Then substitute the value for W in order to solve for L.
24 − W = L
24 − 17 = L
7 = L

300) The correct answer is A. $x = \log_y Z$ is the same as $y^x = Z$, so $4 = \log_4 256$ is the same as $4^4 = 256$.

301) The correct answer is B. y is positive even when x is negative, so we know we are dealing with a squared number. Looking at the value for $x = 5$, we can see that the output for y is 32. This result can be achieved only when 2 is raised to the fifth power. So the correct function is $f(x) = 2^x$.

302) The correct answer is B. $4^{2x} = 64$ is the same as $2x = \log_4 64$. If $4^{2x} = 64$, we need to multiply 4 times itself to get the correct value. $4 \times 4 \times 4 = 64$, which can be expressed as $4^3 = 64$. Our original equation was $4^{2x} = 64$, so to get 3 as an exponent, we need to multiply 2 by 1.5. So, $x = 1.5$.

303) The correct answer is D. Put the values provided for x into the function to solve.
$f_1 = x^2 + x = 5^2 + 5 = 25 + 5 = 30$

304) The correct answer is A. The trig formulas state that $\cos^2 A = 1 − \sin^2 A$. The problem tells us that the sine squared of angle B is 0.1403301, so put the value into the formula to solve.
$\cos^2 B = 1 − \sin^2 B$
$\cos^2 B = 1 − 0.1403301$
$\cos^2 B = 0.8596699$

305) The correct answer is D.
$\cos^2 A = 1 − \sin^2 A$
So, we square sine: $0.1908 \times 0.1908 = 0.036405$.
Then, subtract this result from 1.
$\cos^2 A = 1 − \sin^2 A$
$\cos^2 A = 1 − 0.036405$
$\cos^2 A = 0.963595$

306) The correct answer is D. The street that runs from the gas station to the park (GP) is opposite the courthouse. We have the length of the hypotenuse, so we use sine ("opposite over hypotenuse") to solve.
$\sin 65° = \text{opposite}/\text{hypotenuse}$
$\sin 65° = GP/4$
$4 \times \sin 65° = GP$

307) The correct answer is C. We need to use the formula to calculate the length of the arc: $s = r\theta$
Remember that θ = the radians of the subtended angle, s = arc length, and r = radius. So, substitute values into the formula to solve the problem. In our problem: radius (r) = 16 and radians (θ) = $3\pi/4$
$s = r\theta;\ s = 16 \times 3\pi/4 = 12\pi$

308) The correct answer is C. Substitute the values into the equation to solve. For $x = 2$ and $y = 4$, $5x + 6y = (5 \times 2) + (6 \times 4) = 10 + 24 = 34$.

309) The correct answer is D. The slopes of perpendicular lines are negative reciprocals of each other. The equation for line K is in the slope-intercept form: $y = 5x + 0$, so the slope of line K is 5. To find the reciprocal of line K, you need to invert the whole number (5) to make a fraction. So, 5 becomes $\frac{1}{5}$. You then need to make this a negative number, so $\frac{1}{5}$ becomes $-\frac{1}{5}$. Line L has a y intercept of 0 because the facts of the question state that line L passes through (0, 0). Using the slope intercept formula for line L with a slope of $-\frac{1}{5}$ and a y intercept of 0, we get our answer: $y = -\frac{1}{5}x + 0$. Removing the zero, we get $y = -\frac{1}{5}x$.

310) The correct answer is C. Let's say the number widgets is represented by D and the number whatsits is represented by H. Your equation is: (D × $2) + (H × $25) = $85. We know that the number of whatsits is 3, so put that in the equation and solve for the number of widgets.
(D × $2) + (H × $25) = $85
(D × $2) + (3 × $25) = $85
(D × $2) + $75 = $85
(D × $2) + 75 − 75 = $85 − $75
(D × $2) = $10
$2D = $10
$2D ÷ 2 = $10 ÷ 2
D = 5

311) The correct answer is D. Remember that the domain of a function is all possible x values for the function. You need to avoid any mathematical operations that do not have real number solutions, such as dividing by a zero or finding the square root of a negative number. $f(x) = x \div (1 + x)$, so to avoid dividing by a zero, $x \neq -1$. $g(x) = 1 \div x$, so $x \neq 0$. To find the domain of both functions, put these two results together to state the exclusions to the domain. Therefore, the domain of $f + g$ is all real numbers except 0 and −1.

312) The correct answer is C. The function $(x) = ax^2 + bx + c$ is graphed as a parabola. The other equation is graphed as a non-vertical line that intersects the parabola at two points.

313) The correct answer is B. Remember that $\sqrt{x} = x^{\frac{1}{2}}$, so $\sqrt{3} = 3^{\frac{1}{2}}$

314) The correct answer is A. Miles per hour (MPH) is calculated as follows: miles ÷ hours = MPH. So, if we have the MPH and the miles traveled, we need to change the above equation in order to calculate the hours.
miles ÷ hours = MPH
miles ÷ hours × hours = MPH × hours
miles = MPH × hours
miles ÷ MPH = (MPH × hours) ÷ MPH
miles ÷ MPH = hours

In other words, we divide the number of miles by the miles per hour to get the time for each part of the event. So, for the first part of the event, the hours are calculated as follows: 80 ÷ 5. For the second part of the event, we take the remaining mileage and divide by the unknown variable: 20 ÷ x. Since the event is divided into two parts, these two results added together equal the total time.
Total time = [(80 ÷ 5) + (20 ÷ x)]

The total amount of miles for the event is then divided by the total time to get the average miles per hour for the entire event. We have a 100 mile endurance event, so the result is: 100 ÷ [(80 ÷ 5) + (20 ÷ x)]

315) The correct answer is C. If she uses 12 cups of sugar, she is using 6 times the basic amount of 2 cups. (2 cups × 6 = 12 cups). So, to keep things in proportion, we also need to multiply the basic amount of one-third cup of butter by six: $\frac{1}{3}$ × 6 = 2 cups of butter

316) The correct answer is C. We know that she traveled 150 miles before the repair. Miles traveled before needing the repair: 60 MPH × 2.5 hours = 150 miles traveled. If the journey is 240 miles in total, she has 90 miles remaining after the car is repaired: 240 − 150 = 90. If she then travels at 75 miles an hour for 90 miles, the time she spends is: 90 ÷ 75 = 1.2 hours. There are 60 minutes in an hour, so 1.2 hours is 1 hour and 12 minutes because 60 minutes × 0.20 = 12 minutes. The time spent traveling after the repair is 1 hour and 12 minutes. Now add together all of the times to get your answer: Time spent before needing the repair: 2.5 hours = 2 hours and 30 minutes; Time spent waiting for the repair: 2 hours; The time spent traveling after the repair: 1 hour and 12 minutes; Total time: 5 hours and 42 minutes. If she left home at 6:00 am, she will arrive in Denver at 11:42 am.

317) The correct answer is C. If the amount earned from selling jackets was one-third that of selling jeans, the ratio of jacket to jean sales was 1 to 3. So, we need to divide the total sales of $4,000 into $1,000 for jackets and $3,000 for jeans. We can then solve as follows:
$3,000 in jeans sales ÷ $20 per pair = 150 pairs sold

318) The correct answer is D. Divide each side of the equation by 3. Then subtract 5 from both sides of the equation as shown below.
18 = 3(x + 5)
18 ÷ 3 = [3(x + 5)] ÷ 3
6 = x + 5
6 − 5 = x + 5 − 5
1 = x

319) The correct answer is D. First, we have to calculate the output for our first production method for 10 days: D^5 + 12,000 = 10^5 + 12,000 = 100,000 + 12,000 = 112,000
Then we have to calculate the output for the other production method: 10 × 10,000 = 100,000
112,000 is greater than the 100,000 amount for method B yields.
So, the greatest amount of production for 10 days is 112,000 bottles.

320) The correct answer is A. The tank has a 500 gallon capacity and it is being filled at a rate of 3.5 gallons per minute, so we divide to get the time: 500 ÷ 3.5 = 142.85 minutes ≈ 2 hours and 23 minutes. Remember, of course, that there are 60 minutes in one hour.

Solutions and Explanations for Practice Test Set 5 – Questions 321 to 400

321) The correct answer is C. Count how many blocks lie along the outer edges of the shaded area in order to get your result: Top boundary = 4 feet; Left side boundary = 5 feet; Bottom boundary = 3 feet; Right boundary = 6 feet (Don't forget to count the piece shaped like the upside-down "L" on the right.) Then add these amounts to get your result: 4 + 5 + 3 + 6 = 18 feet.

322) The correct answer is B. First of all, add up the number of questions answered correctly: 12 + 20 + 32 + 32 = 96. Then add up the total number of questions: 15 + 25 + 35 + 45 = 120. Now divide the number of questions answered correctly by the total number of questions to get her percentage: 96 ÷ 120 = 80%

323) The correct answer is D. STEP 1: Convert into minutes the amount of time required to make one cap: 4 hours and 10 minutes = (4 × 60) + 10 = 240 + 10 = 250 minutes needed to make one cap. STEP 2: Multiply by the total output: 250 minutes × 12 caps = 3000 minutes. STEP 3: Convert the total amount of minutes back to hours and minutes: 3000 minutes ÷ 60 = 50 hours.

324) The correct answer is D. Add the feet above ground to the feet below ground to get the total distance: 525 + 95 = 620 feet.

325) The correct answer is C. STEP 1: Add the items together to get the total amount of items available: 13 + 22 + 25 = 60 balloons in total. STEP 2: Divide the amount of items available by the number of people: 60 ÷ 12 = 5.

326) The correct answer is B. STEP 1: Determine the value of the discount by multiplying the normal price by the percentage discount: $90 × 15% = $13.50 discount. STEP 2: Subtract the value of the discount from the normal price to get the new price: $90 − $13.50 = $76.50.

327) The correct answer is C. STEP 1: You can express the fractions as decimals for the sake of simplicity: 10½ = 10.50; 7¾ = 7.75. STEP 2: Then subtract to find the increase: 10.50 − 7.75 = 2.75. STEP 3: Then convert back to a mixed number: 2.75 = 2¾

328) The correct answer is A. After her raise, she earns $184 per week. She continues to work 23 hours per week. STEP 1: Determine the new hourly rate: $184 ÷ 23 hours = $8 per hour. STEP 2: Determine the change in the hourly rate: $8 - $7.50 = 50 cents per hour.

329) The correct answer is A. STEP 1: Determine the distance traveled. If he is traveling 70 miles an hour, he will have traveled 70 miles after one hour has passed. STEP 2: Determine the distance from the towns listed on the sign, considering that he has traveled for one hour. Washington: 140 − 70 = 70 miles from Washington; Yorkville: 105 − 70 = 35 miles from Yorkville; Zorster: 210 − 70 = 140 miles from Zorster. STEP 3: Compare the above figures to your answer choices to get your result. After an hour, he is 70 miles from Washington, so A is correct.

330) The correct answer is D. STEP 1: Subtract the excess from the total: 300 − 114 = 186. STEP 2: Allocate the difference into its respective parts. We are dividing the day into two parts: morning and afternoon. There were 186 cars in total without the excess, so divide this into two parts: 186 ÷ 2 = 93.

STEP 3: Determine the amount for the larger part. There were 114 more cars in the morning, so add this back in: 93 + 114 = 207 cars in the morning.

331) The correct answer is B. STEP 1: Think about the value of the four pairs of socks she is getting in the exchange. These socks cost 50 cents more each than the pairs she has already bought. So, we can express the difference in value of those four pairs of socks as: 4 × ($3 - $2.50). STEP 2: Take into account the value of the extra pair of socks. She paid $2.50 for a fifth pair of socks, but she is only getting four pairs back on the exchange, so she is owed money back for that part of the purchase. Therefore, we can calculate the refund owing as $2.50 – 4($3 - $2.50)

332) The correct answer is A. The line in any fraction can be treated as the division symbol. Accordingly, we can divide by the denominator, which is 100 in this case.

$$\frac{35 \times 90}{100} = (35 \times 90) \div 100$$

333) The correct answer is C. Each journey is 108 minutes (1 hour and 48 minutes) in duration. So, we need to add 108 minutes to the departure time of 11:52 to get the arrival time of 1:40.

334) The correct answer is D. A 12 pound container of item B costs $48. Therefore, it costs $4 per pound ($48 ÷ 12 pounds = $4 per pound). Item C costs 20% more per pound than item B. In other words, Item C costs 80 cents more ($4 × 20% = 0.80). So, the cost per pound of item C is $4.80.

335) The correct answer is B. The cost of the photography course is $20 per week plus a $5 fee per week for review of photographs and administration. So, the course costs $25 per week. To get the total cost we need to multiply by the number of weeks, which is represented by variable W. Therefore, the total cost of the course and fees for W weeks is $25 × W = $25W.

336) The correct answer is A. Remember that when two fractions have the same numerator, the fraction with the smaller number in the denominator is the larger fraction. So, $-1/4$ is less than $1/8$, $1/8$ is less than $1/6$, and $1/6$ is less than 1.

337) The correct answer is C. This question asks you to interpret a graph in order to determine the price per unit of an item. To solve the problem, look at the graph and then divide the total sales in dollars by the total quantity sold in order to get the price per unit. For ten hamburgers, the total price is $85, so each hamburger sells for $8.50: $85 total sales in dollars ÷ 10 hamburgers sold = $8.50 each. The cost of shakes is represented by: $c = \frac{9}{4}s \cdot \frac{9}{4} = (9 \div 4)s = 2.25s$, so each shake costs $2.25. So, the difference between the cost of one hamburger and the cost of one shake is $8.50 – $2.25 = 6.25

338) The correct answer is D. The tenths place is the first place to the right of the decimal, so 12.86749 rounded to the nearest tenth is 12.9. We have to round up because the number in the hundredths place (7) is 5 or greater.

339) The correct answer is C. The distance between point B and point C is 1.2. Point B is at 0.35, so point C is either 0.35 – 1.2 = –0.85 or 0.35 + 1.2 = 1.55.

340) The correct answer is B. 15.845 + 8.21 = 24.055. Rounding to the nearest integer, we remove the decimals to get 24.

341) The correct answer is C. Three out of ten students are taking the class. So, here we have the proportion 3 to 10. STEP 1: Divide the total number of students by the second number in the proportion to get the number of groups: 650 ÷ 10 = 65 groups. STEP 2: Multiply the number of groups by the first number in the proportion in order to get the result: 3 × 65 = 195 art students.

342) The correct answer is A. Put the numbers is ascending order: 2, 2, 3, 5, **6**, **8**, 8, 10, 12, 21. Here, we have got an even number of items, so we need to take an average of the two items in the middle: (8 + 6) ÷ 2 = 7

343) The correct answer is B. This question is asking you to determine the value missing from a sample space when calculating basic probability. This is like other problems on basic probability, but we need to work backwards to find the missing value. First, set up an equation to find the total items in the sample space. Then subtract the quantities of the known subsets from the total in order to determine the missing value. We will use variable T as the total number of items in the set. The probability of getting a red ribbon is $1/3$. So, set up an equation to find the total items in the data set:

$$\frac{5}{T} = \frac{1}{3}$$

$$\frac{5}{T} \times 3 = \frac{1}{3} \times 3$$

$$\frac{5}{T} \times 3 = 1$$

$$\frac{15}{T} = 1$$

$$\frac{15}{T} \times T = 1 \times T$$

$$15 = T$$

We have 5 red ribbons, 6 blue ribbons, and x green ribbons in the data set that make up the total sample space, so now subtract the amount of red and blue ribbons from the total in order to determine the number of green ribbons.
$5 + 6 + x = 15$
$11 + x = 15$
$11 - 11 + x = 15 - 11$
$x = 4$

344) The correct answer is D. The first three bars of the graph represent the first 30 minutes, so add these three amounts together for your answer: 1.5 + 1.2 + 0.8 = 3.5 miles

345) The correct answer is C. The question is asking us how many residents have more than 3 relatives nearby, so we need to add the bars for 4 and 5 relatives from the chart. 20 residents have 4 relatives

nearby and 10 residents have 5 relatives nearby, so 30 residents (20 + 10 = 30) have more than 3 relatives nearby.

346) The correct answer is A. Find the amount of items in the sample set before anything is removed from the set: 4 + 2 + 1 + 4 + 5 = 16. One rope has been removed, so deduct that from the sample space for the second draw: 16 − 1 = 15. The box original had 4 pieces of blue rope, and one piece of blue rope has been removed, so there are 3 pieces of blue rope available for the second draw. So, the probability is $3/15 = 1/5$

347) The correct answer is D. Questions like this one are asking you how to express percentages graphically. Facts such as x students from y total students participate in a group can be represented as x/y. Ten out of 25 students participate in drama club. First of all, express the relationship as a fraction: $10/25$. Then divide to find the percentage: $10/25 = 10 \div 25 = 0.40 = 40\%$. Finally, choose the pie chart that has 40% of its area shaded in dark gray. 40% is slightly less than half, so you need to choose chart D.

348) The correct answer is C. For questions that ask you to interpret bar graphs, you need to read the problem carefully to determine what is represented on the horizontal axis (bottom) and the vertical axis (left side) of the graph. The quantity of diseases is indicated on the bottom of the graph, while the number of children is indicated on the left side of the graph. To determine the amount of children that have been vaccinated against three or more diseases, we need to add the amounts represented by the bars for 3, 4, and 5 diseases: 30 + 20 + 10 = 60 children

349) The correct answer is B. The range is the highest amount minus the lowest amount: 91 − 54 = 37

350) The correct answer is A. What do we do when no number appears in the set more than once? If no number is duplicated, then we say that the data set has no mode.

351) The correct answer is C.
$(5x - 2)(3x^2 + 5x - 8) =$
$(5x \times 3x^2) + (5x \times 5x) + (5x \times -8) + (-2 \times 3x^2) + (-2 \times 5x) + (-2 \times -8) =$
$15x^3 + 25x^2 - 40x - 6x^2 - 10x + 16 =$
$15x^3 + 25x^2 - 6x^2 - 40x - 10x + 16 =$
$15x^3 + 19x^2 - 50x + 16$

352) The correct answer is A. You can subtract the second equation from the first equation as the first step in solving the problem. Look at the term containing x in the second equation. $8x$ is in the second equation. In order to eliminate the term containing x, we need to multiply the first equation by 8.
$x + 5y = 24$
$(x \times 8) + (5y \times 8) = 24 \times 8$
$8x + 40y = 192$

Now subtract.
$8x + 40y = 192$
$-(8x + 2y = 40)$
$38y = 152$

Then solve for y.
$38y = 152$
$38y \div 38 = 152 \div 38$
$y = 4$

Now put the value for y into the first equation and solve for x.
$x + 5y = 24$
$x + (5 \times 4) = 24$
$x + 20 = 24$
$x = 4$
$x = 4$ and $y = 4$, so the answer is (4, 4).

353) The correct answer is C. Find the lowest common denominator. Then add the numerators and simplify.

$$\frac{2}{10x} + \frac{3}{12x^2} =$$

$$\left(\frac{2 \times 6x}{10x \times 6x}\right) + \left(\frac{3 \times 5}{12x^2 \times 5}\right) =$$

$$\frac{12x}{60x^2} + \frac{15}{60x^2} = \frac{12x + 15}{60x^2} =$$

$$\frac{3(4x + 5)}{3 \times 20x^2} = \frac{\cancel{3}(4x + 5)}{\cancel{3} \times 20x^2} = \frac{4x + 5}{20x^2}$$

354) The correct answer is B. Find a perfect square for one of the factors for each radical. Then factor the integers inside each of the square root signs.

$\sqrt{50} + 4\sqrt{32} + 7\sqrt{2} =$
$\sqrt{25 \times 2} + 4\sqrt{16 \times 2} + 7\sqrt{2} =$
$5\sqrt{2} + (4 \times 4)\sqrt{2} + 7\sqrt{2} =$
$5\sqrt{2} + 16\sqrt{2} + 7\sqrt{2} = 28\sqrt{2}$

355) The correct answer is C. First perform the division on the integers: $10 \div 2 = 5$
Then do the division on the other variables.

$a^2 \div a = a$

$b^3 \div b^2 = b$

$c \div c^2 = \frac{1}{c}$

Then multiply these results to get the solution.

$5 \times a \times b \times \frac{1}{c} = \frac{5ab}{c} = 5ab \div c$

356) The correct answer is B. Find the lowest common denominator.

$$\frac{\sqrt{48}}{3} + \frac{5\sqrt{5}}{6} = \left(\frac{\sqrt{48}}{3} \times \frac{2}{2}\right) + \frac{5\sqrt{5}}{6} = \frac{2\sqrt{48}}{6} + \frac{5\sqrt{5}}{6}$$

Then simplify, if possible: $\frac{2\sqrt{48}}{6} + \frac{5\sqrt{5}}{6} = \frac{2\sqrt{(4 \times 4) \times 3}}{6} + \frac{5\sqrt{5}}{6} = \frac{(2 \times 4)\sqrt{3} + 5\sqrt{5}}{6} = \frac{8\sqrt{3} + 5\sqrt{5}}{6}$

357) The correct answer is B. Substitute 1 for x: $\frac{x-3}{2-x} = \frac{1-3}{2-1} = (1 - 3) \div (2 - 1) = -2 \div 1 = -2$

358) The correct answer is D. Multiply the amounts inside the radical sign, but leave the cube root as it is:
$\sqrt[3]{5} \times \sqrt[3]{7} = \sqrt[3]{35}$

359) The correct answer is D. Simplify the numerator and multiply the radicals in the denominator using the FOIL method. Then simplify the denominator.
$\frac{1}{\sqrt{x} - \sqrt{y}} \times \frac{\sqrt{x} + \sqrt{y}}{\sqrt{x} + \sqrt{y}} = \frac{\sqrt{x} + \sqrt{y}}{\sqrt{x}^2 + \sqrt{xy} - \sqrt{xy} - \sqrt{y}^2} = \frac{\sqrt{x} + \sqrt{y}}{\sqrt{x}^2 - \sqrt{y}^2} = \frac{\sqrt{x} + \sqrt{y}}{x - y}$

360) The correct answer is D. Factor each of the parentheticals in the expression: $(3x + 3y)(5a + 5b) = 3(x + y) \times 5(a + b)$. We know that $x + y = 5$ and $a + b = 4$, so we can substitute the values for each of the parentheticals: $3(x + y) \times 5(a + b) = 3(5) \times 5(4) = 15 \times 20 = 300$

361) The correct answer is D.
Step 1: Factor the equation.
$x^2 + 6x + 8 = 0$
$(x + 2)(x + 4) = 0$
Step 2: Now substitute 0 for x in the first pair of parentheses.
$(0 + 2)(x + 4) = 0$
$2(x + 4) = 0$
$2x + 8 = 0$
$2x + 8 - 8 = 0 - 8$
$2x = -8$
$2x \div 2 = -8 \div 2$
$x = -4$
Step 3: Then substitute 0 for x in the second pair of parentheses.
$(x + 2)(x + 4) = 0$
$(x + 2)(0 + 4) = 0$
$(x + 2)4 = 0$
$4x + 8 = 0$
$4x + 8 - 8 = 0 - 8$
$4x = -8$
$4x \div 4 = -8 \div 4$
$x = -2$

362) The correct answer is C. The line in a fraction is the same as the division symbol. For example, $^a/_b = a \div b$. In the same way, $^3/_{xy} = 3 \div (xy)$.

363) The correct answer is D. Get the integers to one side of the equation first of all.
$\frac{1}{5}x + 3 = 5$

$$\frac{1}{5}x + 3 - 3 = 5 - 3$$

$$\frac{1}{5}x = 2$$

Then multiply to eliminate the fraction and solve the problem.

$$\frac{1}{5}x \times 5 = 2 \times 5$$

$$x = 10$$

364) The correct answer is C.
Factor: $x^2 - 12x + 35 < 0$
$(x - 7)(x - 5) < 0$
Then solve each parenthetical for zero:
$(x - 7) = 0$
$7 - 7 = 0$
$x = 7$

$(x - 5) = 0$
$5 - 5 = 0$
$x = 5$
So, $5 < x < 7$

Now check. Use 6 to check to for $x < 7$. Since $6 < 7$ is correct, our proof for this should also be correct.
$x^2 - 12x + 35 < 0$
$6^2 - (12 \times 6) + 35 < 0$
$36 - 72 + 35 < 0$
$-36 + 35 < 0$
$-1 < 0$ CORRECT

Use 4 to check for $x > 5$, which is the same as $5 < x$. Since $4 > 5$ is incorrect, our proof for this should be incorrect.
$x^2 - 12x + 35 < 0$
$4^2 - (12 \times 4) + 35 < 0$
$16 - 48 + 35 < 0$
$-32 + 35 < 0$
$3 < 0$ INCORRECT

So, we have proved that $5 < x < 7$.

365) The correct answer is C. Each term contains the variables x and y. So, factor out *xy* as shown:
$2xy - 8x^2y + 6y^2x^2 = xy(2 - 8x + 6xy)$. Then, factor out any whole numbers. All of the terms inside the parentheses are divisible by 2, so factor out 2: $xy(2 - 8x + 6xy) = 2xy(1 - 4x + 3xy)$

366) The correct answer is D. When looking at scatterplots, try to see if the dots are roughly grouped into any kind of pattern or line. If so, positive or negative relationships may be represented. Here, however,

the dots are located at what appear to be random places on all four quadrants of the graph. So, the scatterplot suggests that there is no relationship between *x* and *y*.

367) The correct answer is C. The formula for circumference is: $\pi \times 2 \times R$. The center of the circle is on (0, 0) and the top edge of the circle extends to (0, 3), so the radius of the circle is 3. Therefore, the circumference is: $\pi \times 2 \times 3 = 6\pi$

368) The correct answer is B. The interior of the garage is square, so its volume is calculated by taking the length times the width times the height: $20 \times 20 \times 18 = 7200$. The roof of the garage is a pyramid shape, so its volume is calculated by taking one-third of the base length squared times the height: $(20 \times 20 \times 15) \times 1/3 = 6000 \div 3 = 2000$. The volume of the roof to the that of the interior is: $\frac{2000}{7200} = \frac{5}{18}$

369) The correct answer is B. Circumference is $2\pi R$, so the circumference of the large wheel is 20π and the circumference of the smaller wheel is 12π. If the large wheel travels 360 revolutions, it travels a distance of: $20\pi \times 360 = 7200\pi$. To determine the number of revolutions the small wheel needs to make to go the same distance, we divide the distance by the circumference of the smaller wheel: $7200\pi \div 12\pi = 600$. Finally, calculate the difference in the number of revolutions: $600 - 360 = 240$

370) The correct answer is A. The diameter of the circle is 2, so the circumference is 2π. There are 360 degrees in a circle and the question is asking us about a 20 degree angle, so the arc length relates to one-eighteenth of the circumference: $20 \div 360 = 1/18$. So, we need to take one-eighteenth of the circumference to get the answer: $2\pi \times 1/18 = 2\pi/18 = \pi/9$

371) The correct answer is B. The two sides of the field form a right angle, so we can use the Pythagorean theorem to solve the problem: $\sqrt{3^2 + 4^2} = \sqrt{9 + 16} = \sqrt{25} = 5$

372) The correct answer is A. Since the left and right sides of this figure are not parallel, the figure is classified as a trapezoid. To find the area of a trapezoid we take the average of the length of the top (T) and bottom (B) and multiply by the height (H):
$\frac{T+B}{2} \times H = \frac{7+12}{2} \times 6 = 9.5 \times 6 = 57$

373) The correct answer is D. To find the area of the rectangle, we must first find the missing length of the one side by subtracting: 15 − 6 = 9. Then multiply to find the area of the rectangle: 9 × 5 = 45. Then find the area of the triangle: (6 × 5) ÷ 2 = 15. Finally, add to find the total area: 45 + 15 = 60.

374) The correct answer is C. The answer is the radian equation for 90 degrees: $\pi \div 2 \times \text{radian} = 90°$

375) The correct answer is C. We simply divide to get the answer: 64 ÷ 4 = 16

376) The correct answers is B. A parallelogram is a four-sided figure that has two pairs of parallel sides. The opposite or facing sides of a parallelogram are of equal length and the opposite angles of a parallelogram are of equal measure. You will recall that congruent is another word for equal in measure. So, answer B is correct. A rectangle is a parallelogram with four angles of equal size (all of which are right angles), while a square is a parallelogram with four sides of equal length and four right angles.

377) The correct answer is B. Two angles are supplementary if they add up to 180 degrees.

378) The correct answer is A. The problem tells us that A is 3 times B, and B is 3 more than 6 times C. So, we need to create equations based on this information.
B is 3 more than 6 times C: B = 6C + 3
A is 3 times B: A = 3B
Since B = 6C + 3, we can substitute 6C + 3 for B in the second equation as follows:
A = 3B
A = 3(6C + 3)
A = 18C + 9
So, A is 9 more than 18 times C.

379) The correct answer is C. Looking at the value for x = 100, we can see that the output for y is 10. We can get this result only when the output is the square root of x. so the correct function is $f(x) = \sqrt{x}$

380) The correct answer is B. Perform the operation inside the absolute value signs: 6 – 13 = –7. The absolute value of –7 is 7.

381) The correct answer is B. y^x = Z is the same as x = $\log_y Z$, so 9^2 = 81 is the same as 2 = $\log_9 81$.

382) The correct answer is B. Put the values provided for x into the second function. $f_2(9) = \sqrt{9} + 3 = 3 + 3 = 6$. Then put this result into the first function. $f_1(6) = 3 \times 6 + 1 = 19$

383) The correct answer is C. Line segment YG is the opposite side of this triangle, which is the unknown for our problem. We have the length of the adjacent side, which is side BG, with a length of 2 miles. So, we need to use "opposite over adjacent" or the tangent of 60 degrees to solve the problem.
tan = opposite/adjacent
tan 60° = YG ÷ BG
tan 60° = YG ÷ 2 miles
2 miles × tan 60° = YG ÷ 2 miles × 2 miles
2 miles × tan 60° = YG

384) The correct answer is B. According to the trig formulas, sin A = cos(90° – A). This is a 72 degree angle, so sin 72 = cos(90 – 72) = cos 18. The sine of a 72 degree angle is 0.951056516, and the cosine of an 18 degree angle is also 0.951056516.

385) The correct answer is A. The trig formulas state: tan ϴ = sin ϴ ÷ cos ϴ and tan ϴ × cos ϴ = sin ϴ. The cosine of A is 0.70710678 and the tangent of angle A is 1, so put the values into the formula.
tan ϴ × cos ϴ = sin ϴ
1 × 0.70710678 = 0.70710678
So, sin A = 0.70710678
You can also solve this simply by knowing that cosine and sine are the same for a 45 degree angle.

386) The correct answer is B. From the radian formulas in part 1, we know that $\pi \div 2 \times$ radian = 90°

387) The correct answer is D. Our points are (5, 7) and (11, –3) so use the midpoint formula.

$(x_1 + x_2) \div 2$, $(y_1 + y_2) \div 2$
$(5 + 11) \div 2$ = midpoint x, $(7 - 3) \div 2$ = midpoint y
$16 \div 2$ = midpoint x, $4 \div 2$ = midpoint y
8 = midpoint x, 2 = midpoint y

388) The correct answer is C. In order to find the value of a variable inside a square root sign, you need to square each side of the equation.
$\sqrt{9z + 18} = 9$
$\sqrt{9z + 18}^2 = 9^2$
$9z + 18 = 81$
$9z + 18 - 18 = 81 - 18$
$9z = 63$
$9z \div 9 = 63 \div 9$
$z = 7$

389) The correct answer is C. First you need to get rid of the fraction. To eliminate the fraction, multiply each side of the equation by the denominator of the fraction.
$z = \dfrac{x}{1 - y}$
$z \times (1 - y) = \dfrac{x}{1 - y} \times (1 - y)$
$z(1 - y) = x$

Then isolate y to solve.
$z(1 - y) \div z = x \div z$
$1 - y = x \div z$
$1 - 1 - y = (x \div z) - 1$
$-y = (x \div z) - 1$
$-y \times -1 = [(x \div z) - 1] \times -1$
$y = -\dfrac{x}{z} + 1$

390) The correct answer is D. Multiply the radical in front of the parentheses by each radical inside the parentheses. Then simplify further if possible.
$\sqrt{6} \cdot (\sqrt{40} + \sqrt{6}) =$
$(\sqrt{6} \times \sqrt{40}) + (\sqrt{6} \times \sqrt{6}) =$
$\sqrt{240} + 6 = \sqrt{16 \times 15} + 6 = 4\sqrt{15} + 6$

391) The correct answer is C. When you have fractions as exponents, the denominator of the faction is placed in front of the radical sign. The numerators become the new exponents: $a^{1/2} b^{1/4} c^{3/4} =$
$\sqrt{a} \times \sqrt[4]{b} \times \sqrt[4]{c^3}$

392) The correct answer is A. If the base is the same, and you need to divide, you subtract the exponents: $ab^8 \div ab^2 = ab^{8-2} = ab^6$

393) The correct answer is A. The line $y = 5$ is horizontal and parallel to the x axis. So, the point of intersection with the circle also needs to have a y coordinate of 5. Coordinates (0, 5) meet this criteria.

394) The correct answer is C. He is using 14 ounces of plaster powder, so divide that by the basic amount of plaster of 4 ounces, stated in the instructions. $14 \div 4 = 3.5$. Then multiply this result by the basic amount of water of 3 ounces, stated in the instructions: $3 \times 3.5 = 10.5$

395) The correct answer is C. The y coordinate where $x = 10$ is $y = 65$. So, the graph represents the function $f(x) = x \times 6.50$. In other words, Item A sells for $6.50 a pound. The equation C = p × (7 ÷ 2) is equal to C = 3.50p because $7 \div 2 = 3.5$. So, Item B sells for $3.50 per pound. Therefore, Item B costs $3 less per pound than Item A.

396) The correct answer is A. The function must be a quadratic because the graph is a parabola. We need a negative constant because y is less than zero at the vertex of the parabola. Answer A is the only answer that has a squared number, making it a quadratic, and a negative constant.

397) The correct answer is C. Any polynomial function of odd degree with a leading negative coefficient will result in negative end behaviour for large positive values of x and positive end behaviour for large negative values of x. For the polynomial to be of odd degree, w must be negative and z must be even.

398) The correct answer is D. Find the lowest common denominator.

$$\frac{1}{a+1} + \frac{1}{a} =$$

$$\left(\frac{1}{a+1} \times \frac{a}{a}\right) + \left(\frac{1}{a} \times \frac{a+1}{a+1}\right) =$$

$$\frac{a}{a^2 + a} + \frac{a+1}{a^2 + a}$$

Then simplify, if possible

$$\frac{a}{a^2 + a} + \frac{a+1}{a^2 + a} =$$

$$\frac{a + a + 1}{a^2 + a} = \frac{2a + 1}{a^2 + a}$$

399) The correct answer is D. Treat the main fraction as the division sign.

$$\frac{5x}{1/xy} = 5x \div \frac{1}{xy}$$

Then invert the second fraction and multiply as usual.

$$5x \div \frac{1}{xy} = 5x \times \frac{xy}{1} = 5x \times xy = 5x^2 y$$

400) The correct answer is B. Convert the seconds to hours: 108,000 seconds ÷ 60 seconds per minute ÷ 60 minutes per hour = 30 hours. Then multiply by the speed of the rocket to get the miles: 25,000 miles per hour × 30 hours = 750,000 miles to travel. Finally, express your answer in exponential form: $750,000 = 7.5 \times 100,000 = 7.5 \times 10^5$

ANSWER KEY

1) D	36) B	71) D	106) A	141) C
2) D	37) B	72) C	107) D	142) C
3) C	38) D	73) A	108) D	143) B
4) B	39) B	74) C	109) B	144) B
5) A	40) C	75) C	110) D	145) C
6) D	41) C	76) A	111) B	146) D
7) C	42) D	77) A	112) B	147) A
8) A	43) A	78) B	113) A	148) C
9) B	44) B	79) B	114) A	149) D
10) C	45) D	80) A	115) C	150) B
11) B	46) C	81) B	116) A	151) D
12) D	47) C	82) B	117) B	152) B
13) B	48) D	83) A	118) C	153) C
14) C	49) A	84) A	119) B	154) C
15) A	50) D	85) C	120) D	155) D
16) A	51) B	86) A	121) D	156) C
17) B	52) D	87) A	122) D	157) D
18) D	53) C	88) D	123) C	158) A
19) C	54) D	89) A	124) B	159) B
20) D	55) A	90) C	125) C	160) D
21) B	56) B	91) D	126) D	161) C
22) A	57) C	92) D	127) D	162) C
23) D	58) B	93) D	128) C	163) A
24) A	59) D	94) C	129) A	164) C
25) B	60) A	95) C	130) C	165) D
26) C	61) D	96) B	131) D	166) A
27) B	62) C	97) D	132) C	167) D
28) C	63) D	98) C	133) D	168) B
29) C	64) C	99) A	134) D	169) D
30) C	65) C	100) B	135) D	170) D
31) D	66) C	101) B	136) C	171) C
32) D	67) B	102) D	137) B	172) D
33) A	68) D	103) A	138) A	173) A
34) B	69) A	104) D	139) A	174) B
35) A	70) A	105) C	140) D	175) D

176) B	212) D	248) C	284) D	320) A
177) D	213) D	249) B	285) B	321) C
178) B	214) B	250) C	286) C	322) B
179) A	215) B	251) B	287) A	323) D
180) D	216) B	252) D	288) A	324) D
181) B	217) C	253) D	289) D	325) C
182) C	218) A	254) D	290) B	326) B
183) D	219) D	255) C	291) C	327) C
184) B	220) C	256) C	292) A	328) A
185) C	221) D	257) D	293) C	329) A
186) C	222) A	258) D	294) C	330) D
187) B	223) A	259) D	295) A	331) B
188) B	224) D	260) B	296) D	332) A
189) A	225) C	261) C	297) C	333) C
190) C	226) D	262) D	298) B	334) D
191) A	227) B	263) A	299) A	335) B
192) C	228) B	264) A	300) A	336) A
193) D	229) D	265) D	301) B	337) C
194) D	230) D	266) D	302) B	338) D
195) A	231) B	267) C	303) D	339) C
196) D	232) C	268) D	304) A	340) B
197) A	233) B	269) A	305) D	341) C
198) B	234) C	270) B	306) D	342) A
199) B	235) B	271) A	307) C	343) B
200) B	236) A	272) C	308) C	344) D
201) A	237) C	273) B	309) D	345) C
202) D	238) D	274) A	310) C	346) A
203) A	239) B	275) A	311) D	347) D
204) A	240) D	276) C	312) C	348) C
205) C	241) C	277) B	313) B	349) B
206) D	242) D	278) D	314) A	350) A
207) A	243) C	279) C	315) C	351) C
208) C	244) B	280) A	316) C	352) A
209) B	245) D	281) D	317) C	353) C
210) B	246) B	282) B	318) D	354) B
211) D	247) B	283) D	319) D	355) C

356) B	365) C	374) C	383) C	392) A
357) B	366) D	375) C	384) B	393) A
358) D	367) C	376) B	385) A	394) C
359) D	368) B	377) B	386) B	395) C
360) D	369) B	378) A	387) D	396) A
361) D	370) A	379) C	388) C	397) C
362) C	371) B	380) B	389) C	398) D
363) D	372) A	381) B	390) D	399) D
364) C	373) D	382) B	391) C	400) B

www.ingramcontent.com/pod-product-compliance
Lightning Source LLC
Chambersburg PA
CBHW081748100526
44592CB00015B/2335